ROOFING
FLASHING & WATERPROOFING

FROM THE EDITORS OF **Fine Homebuilding**®

The Taunton Press

The Taunton Press
Inspiration for hands-on living®

The Taunton Press, Inc., 63 South Main Street, PO Box 5506, Newtown, CT 06470-5506
e-mail: tp@taunton.com

Jacket/Cover Design: Cathy Cassidy
Interior Design: Cathy Cassidy
Layout: Cathy Cassidy
Front Cover Photographer: Roe A. Osborn, courtesy *Fine Homebuilding*, © The Taunton Press, Inc.
Back Cover Photographers: (clockwise from top left) Steve Culpepper, courtesy *Fine Homebuilding*, © The Taunton Press, Inc.; Tom O'Brien, courtesy *Fine Homebuilding*, © The Taunton Press, Inc.; Roe A. Osborn, courtesy *Fine Homebuilding*, © The Taunton Press, Inc.; Rich Ziegner, courtesy *Fine Homebuilding*, © The Taunton Press, Inc.

Library of Congress Cataloging-in-Publication Data
Roofing, flashing & waterproofing / the editors of Fine homebuilding.
p. cm. -- (For pros, by pros)
Includes index.
ISBN 1-56158-778-8
1. Roofing. 2. Flashing (Building materials) 3. Waterproofing. I. Title: Roofing, flashing, and waterproofing.
II. Fine homebuilding. III. Taunton Press. IV. Series.
TH2430.R64 2005
695--dc22

2005010637

Printed in the United States of America
10 9 8 7 6 5 4 3 2 1

Special thanks to the authors, editors, art directors,
copy editors, and other staff members of *Fine Homebuilding*
who contributed to the development of the articles in this book.

CONTENTS

PART 2: ROOFING

INTRODUCTION

Like a good hat, a roof is supposed to keep your head dry...or cool or warm, depending on the weather. But unlike a hat, a roof also has to protect your family, friends, and worldly possessions, not to mention the very structure of your house itself. That's a lot of responsibility for a job generally entrusted to shirtless twenty-year-olds, working to the blare of rock music and dreaming of their next cold beer. Believe me, I know because I used to be one of those guys.

And when you're up on a roof, clinging to some narrow perch, with your feet bent backwards on themselves, it is always too hot, or too cold, or too wet, and you really don't want to be up there. Under those circumstances, it's hard for even the most conscientious person to care whether the shingle nails are landing precisely in the nailing strip and each piece of step flashing is overlapping the previous one by a minimum of two inches. But care you must, or else water will seep insidiously, like cancer cells, into the bones of the house.

But caring only gets you so far. You also have to know where the shingle nails go and how the step flashing should overlap, along with a thousand other details that go into a good roof. And that's where this book can help. Collected here are 17 articles from past issues of *Fine Homebuilding* magazine. Written by professional builders, who have all done time on the roof, these articles will help you put a sound hat on someone's biggest investment.

—Kevin Ireton,
editor, *Fine Homebuilding*

Flashing Walls

■ BY SCOTT McBRIDE

As a carpenter for more than 20 years, I've seen the mischief water can do when it gets under a building's skin. I've also studied ways to prevent this situation from happening. What I've learned is that water can be coaxed and persuaded to remain on the outside of a building, where it belongs, if you know the right methods and use the right materials. Although these methods and materials are similar for sidewall and roof flashings, this article will look specifically at flashing the various sidewall trouble spots.

Flashings are membranes woven into a structure's exterior cladding at key points to keep water moving out and down (see the photo at left). They work primarily by means of gravity. The underlying principle, therefore, in all flashing, siding, and roofing work is that which is above overlaps that which is below. In addition, well-designed flashing should break the surface tension of water, which will help prevent moisture migration along cracks and between materials.

Yes, some windows can be flashed with tarpaper and housewrap. Lapping asphalt-felt paper and housewrap over the nailing flange of this clad window will help keep water out of the framing.

Copper and Lead Are Traditional Choices

Metal has long been used for flashing because it can be beaten or rolled into thin sheets. Copper in particular has been a popular choice because of its excellent corrosion resistance, even in the presence of an alkaline material such as concrete. It is also strong enough to hold a shape yet soft enough to work easily. Finally, copper can be easily and permanently joined by soldering. This trait makes it possible to fabricate complex, watertight shapes that can't be made by means of bending alone, which is particularly important in roof flashing. Copper for flashing is usually sold in a 16-oz. weight (1 sq. ft. weighs 16 oz.) and is available in soft and hard tempers. Hard, or cold-rolled, copper is stronger and is preferred for most flashing work.

Copper is often used for its traditional appearance, and as it weathers, it develops an attractive, protective surface, or patina. Lead-coated copper has even better corrosion resistance than plain copper, with a price tag to match, and is used primarily in urban environments where air pollution can corrode plain copper prematurely.

Despite its advantages, however, copper does have a few drawbacks. Runoff from plain copper can sometimes cause a greenish staining, although this problem does not occur with lead-coated copper. Another problem is the incompatibility of copper and cedar: The extractives from red-cedar shingles and shakes will deteriorate copper and, therefore, make it a poor choice where red cedar is present. Copper's main disadvantage is price; its cost is three to four times that of aluminum.

Like copper, lead has a centuries-long tradition as a flashing material. It has good corrosion resistance and is highly malleable. It's still used occasionally for custom work where a flashing must be bent to conform to an irregular surface, such as a tile roof. On the downside, lead's softness makes it vulnerable to punctures, and its low melting point makes it difficult to solder; the lead sheet will melt as readily as the solder.

Aluminum Is a Good Choice for Simple Flashings

Most residential flashing work these days is done with aluminum. It is easy to work, is inexpensive, and has good corrosion resistance in most environments. It also readily accepts paint, which can be applied in the field or, better yet, in a factory. However, the paper-thin aluminum that's sold in rolls at lumberyards is barely adequate for flashing work. A better choice is the prefinished 0.029-in. coil stock sold by aluminum-siding distributors. It often comes painted brown on one side and white on the other. The white side makes layout with a pencil easy.

Unfortunately, aluminum can't be soldered easily, so it's limited to situations where simple overlapping will keep water at bay. Caulks and sealants can be used on aluminum, but they often fail due to the metal's high degree of thermal expansion. As

Galvanic Corrosion Can Deteriorate Flashing

An important consideration in the use of metal flashings and fasteners is galvanic corrosion. This reaction occurs when dissimilar metals in the galvanic series—a ranking of a metal's tendency to react with other metals in a specific environment—are brought into contact. For this reason, it's best to keep all metal components in a roof/sidewall system the same, or at least as close as possible to each other within the galvanic series. That includes all metal roofing, roof and sidewall flashings, nails, gutters, and downspouts.

A sheet-metal brake speeds up flashing work. After he checks the measurement at either end of the metal strip, the worker lowers the clamp-bar lever held in his right hand and bends the flashing to the proper angle. An alternative to this portable brake is the site-built one shown in the drawing on the facing page.

the metal stretches and contracts, its bond with the sealant will often break.

Tin-plated steel (terne) and zinc-plated steel (galvanized) can also be used for flashing, though their corrosion resistance is limited without periodic painting. Flashings, though, can't be fully painted after installation because they're hidden behind other materials. Still, galvanized flashing is preferred over aluminum flashing in coastal areas where salt air is particularly corrosive to aluminum.

Do You Need a Brake?

For bending flashings, a portable brake is hard to beat (see the photo above). It's one of the most timesaving tools I know of. My 8-ft. aluminum J-Brake (Van Mark Products Corp., 24145 Industrial Park Dr., Farmington Hills, MI 48335; 800-VAN-MARK) has served me for a decade with no more attention than the occasional dab of grease. (Van Mark no longer carries the J-Brake, but the company sells similar brakes with different names.)

Operating a brake is simplicity itself: stick in the sheet of metal, throw the clamp bar, and lift the brake bar. Voilà! A precise bend. If you need to make a custom flashing but don't have your own brake, call sheet-metal suppliers. Some places keep an old shop brake in a corner of the warehouse for their customers' use.

A site-built brake, made with three equal lengths of 2x6, is an inexpensive alternative (see the top drawing on the facing page) and will produce crisp bends. You can also improvise with a 2x4 straightedge and mallet. Use light blows, going back and forth along the length of the bend several times until the desired angle is reached.

To make small bends and to turn up tabs, I use hand seamers. This tool is a pair of pliers with wide, flat jaws. While I generally use the straight-handled type (Malco® Products Inc., 14080 State Hwy 55 NW, Annandale, MN 55302-0400; 800-596-3494), I also have a bent-handle version, another compound-leverage style that is handy in certain tight spots. I use several pairs of spring-locking seamers for clamping assemblies during soldering. Deep-throat locking pliers with C-shaped jaws are also good for this process.

Cutting metal doesn't require a big investment in tools, either. I do most of my straight cutting with plain tin snips, which are easy to sharpen. For curve cutting, I use aviation snips. Originally developed for the aircraft industry, aviators are maneuverable, and compound action gives them plenty of crunch. Aviation snips are available in right-hand curve, left-hand curve, and straight (traditionally color-coded red, green, and yellow, respectively). Mostly, I use the right-hand curve.

Aluminum flashing is generally cut with a utility knife, a process known as slitting. After it is scored, the aluminum is folded back and forth until it breaks. With a sharp knife, only one or two wags of the metal are necessary to snap the pieces apart.

A Site-Built Sheet-Metal Brake

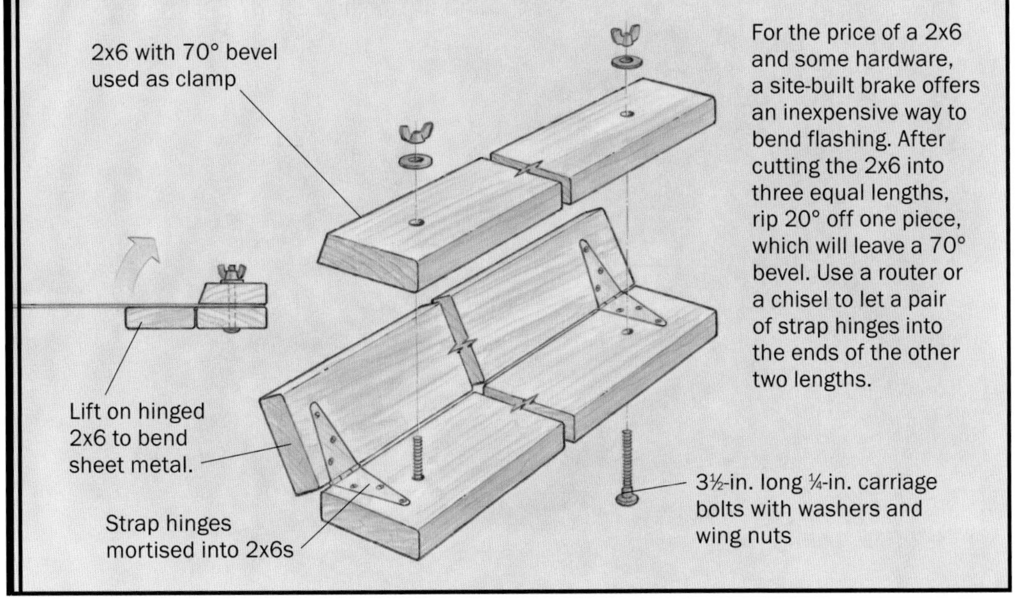

For the price of a 2x6 and some hardware, a site-built brake offers an inexpensive way to bend flashing. After cutting the 2x6 into three equal lengths, rip 20° off one piece, which will leave a 70° bevel. Use a router or a chisel to let a pair of strap hinges into the ends of the other two lengths.

For soldering copper flashing, I use a soldering iron heated continuously by a small acetylene flame. The hose and tank are a nuisance, but I don't worry about the iron cooling off. The more traditional apparatus for soldering is two or more irons heated in a brazier. While one iron is in use, the other is heating in the flames. Braziers can be fueled with charcoal or propane.

Flashing Keeps the Termites Out of the Building

Where required by code, the first flashing to go in a typical house is the termite shield (see the drawing at right). It prevents termites that travel through masonry foundations from getting into wood framing. For termite shield, I use thin aluminum-roll flashing that's 2 in. wider than the foundation wall. I simply roll out the flashing on top of the foundation, beating it down over the anchor bolts as I go, using a mallet or my hammer and a scrap of wood. The wood protects the anchor-bolt threads and is kinder to the aluminum than my framing hammer would be. At corners and splices, I give the aluminum 1 ft. of overlap and caulk between the layers. Then I go back and beat the overhanging metal down over the sides of the foundation. It looks rough, but it eventually gets covered by the sheathing and siding.

Termite Shield

A simple aluminum flashing helps to keep termites from migrating up a foundation wall and into the framing of a house.

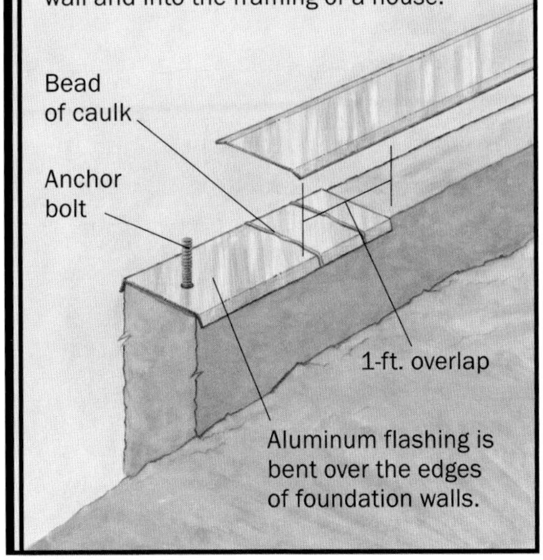

Everything converges at the water table. Flashing keeps water from sneaking in behind the water table; it tucks under the housewrap and the asphalt-felt door splines.

Flashing a Water Table

Some houses, especially older ones, have wide skirtboards between the foundation and the siding. The skirtboard is capped by a sort of sloped sill known as a water table. I flash over the water table with a simple dog-leg flashing before starting the siding (see the photo above). Prefabricated drip cap from the lumberyard may work, or the dimensions may require that I custom-bend flashing on my brake.

To make such a flashing, I first cut off 8-ft. lengths of copper or aluminum roll (the maximum capacity of my brake). I then cut the 8-ft. lengths into strips about 3 in. wide so that the finished flashing can follow the water table profile and extend up the wall at least 1½ in. At both ends of each strip, I punch prick marks with an awl to indicate the fold lines. Prick marks are precise, and they can be read from either side of the sheet of metal, an advantage when bending metal because a piece often needs to be flipped upside down in the brake to fold the right way. For simple flashings, the layout step can be avoided by simply measuring the distance that the metal protrudes beyond the brake.

After marking out, I fold each piece in the brake, eyeballing the degree of bend to approximate the angle between the wall and the sloped water table. It's better to underbend than to overbend; that way, the siding will spring the flashing down tight. I generally overlap flashing sections about 6 in.

Porch and Deck Ledgers Can Be Traps for Water

Where a porch or a deck connects to a house, flashing is used to keep water from running down between the ledger board and the house's band joist. In the case of a tightly laid porch floor, I run a simple L-flashing up under the siding about 4 in. and out over the floor about 3 in. (see the left drawing on the facing page). Because such floors are pitched outward, water runs away from the flashing. To prevent capillary action from sucking moisture back up between the floor and the flashing, I run a bead of caulk between the two. The caulk also helps to hold the flashing tight to the floor.

For exposed decks I also use an L-flashing, but the lower flange runs between the

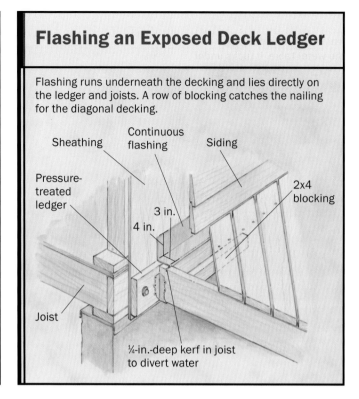

Flashing a Tightly Laid Porch Floor

Flashing extends out and over this porch floor, which is blind-nailed, ¾-in. tongue-and-groove flooring.

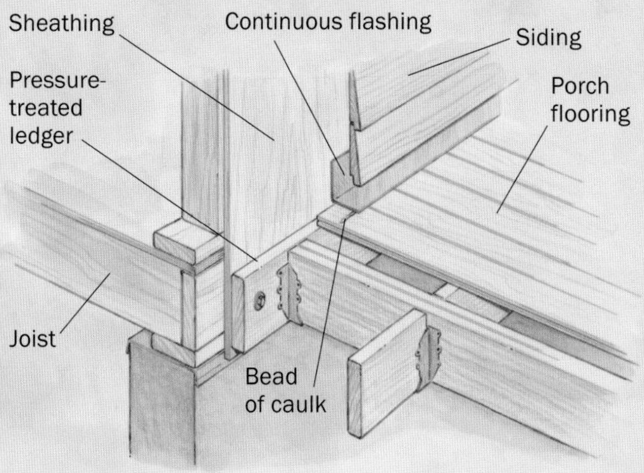

Flashing tucks up behind siding 4 in. and extends 3 in. out over flooring.

Flashing an Exposed Deck Ledger

Flashing runs underneath the decking and lies directly on the ledger and joists. A row of blocking catches the nailing for the diagonal decking.

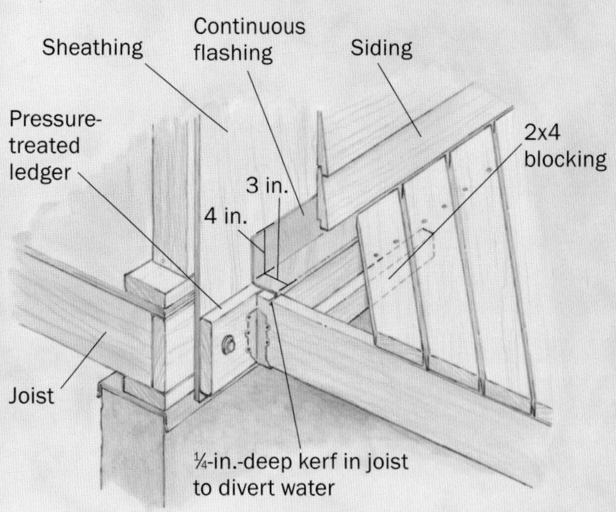

decking and the joist, not on top (see the right drawing above). I lay the deck boards diagonally with their ends cantilevering past a row of blocking set in about 4 in. from the ledger. The decking covers the flashing and is nailed only into the blocking because nailing directly through the flashing into the ledger board would defeat the flashing's purpose. The diagonal decking also adds enormously to the stiffness of the structure.

To prevent water from running back between the joist and the flashing, I make a ¼-in.-deep chainsaw kerf in the top of each joist before the flashing goes on. This kerf catches the seepage and diverts it down over the side of the joist. If kerfing the joist makes you nervous, you can caulk instead.

Asphalt-Felt Splines Keep the Water Out of Openings

It's common practice today to install windows directly over housewrap, with no felt flashing around the sides or bottom. The problem is that if the caulking between the siding and the window fails, water seeps in.

The old-fashioned way of flashing around windows with felt splines is still best. It's done while installing the windows, before the housewrap goes on. The 6-in.-wide splines are cut from rolls of 30-lb. felt and installed as shown in the drawings on p. 10.

Next, the metal head flashing goes on. This flashing can be installed directly over the window's exterior head casing (see the top left drawing on p. 10), or a wooden drip cap can be mounted between the flashing and the casing (see the top center drawing on p. 10). The ends of the head flashing should be turned down over the end grain of the wooden drip cap to make a neat finish (see the bottom drawings on p. 10). With windows that are flanged, the head flashing is not necessary (see the top right drawing on p. 10).

After the head flashing is installed, the housewrap goes on. It runs under the bottom spline and over the head flashing. The housewrap can go over or under the side splines, as long as there is a good overlap.

Weatherproofing Windows and Doors With Splines and Flashing

WOOD-CASED WINDOW SHOULD HAVE A METAL DRIP CAP

Prefabricated or custom-bent metal flashing laps over the window casing and extends up the wall. Splines and housewrap then overlap this drip cap.

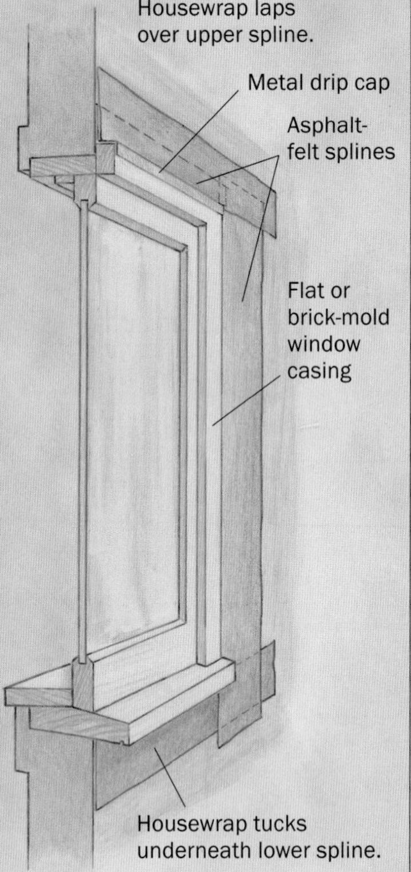

WINDOWS WITH WOODEN DRIP CAPS ALSO NEED METAL FLASHING

The L-flashing laps onto the wooden drip cap and extends up the wall. An asphalt spline laps over this flashing and is overlapped by housewrap.

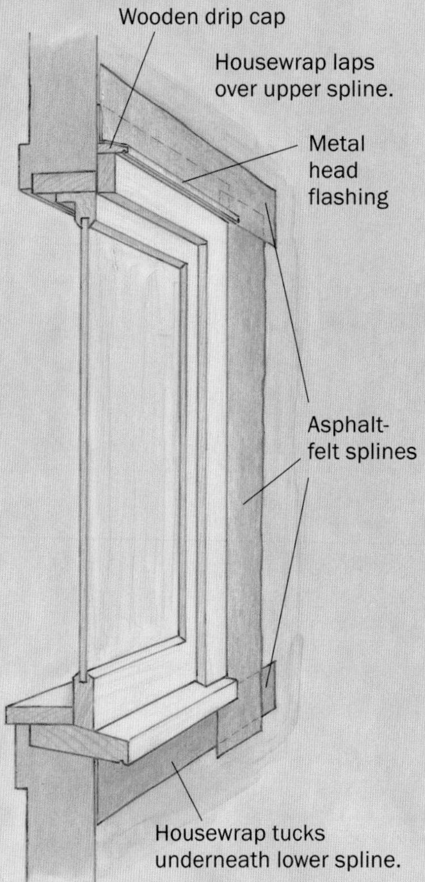

CLAD WINDOWS DON'T NEED METAL HEAD FLASHING

The window's integral nailing flange acts like a flashing. The top flange should be overlapped by an asphalt spline; housewrap laps over top spline and flange.

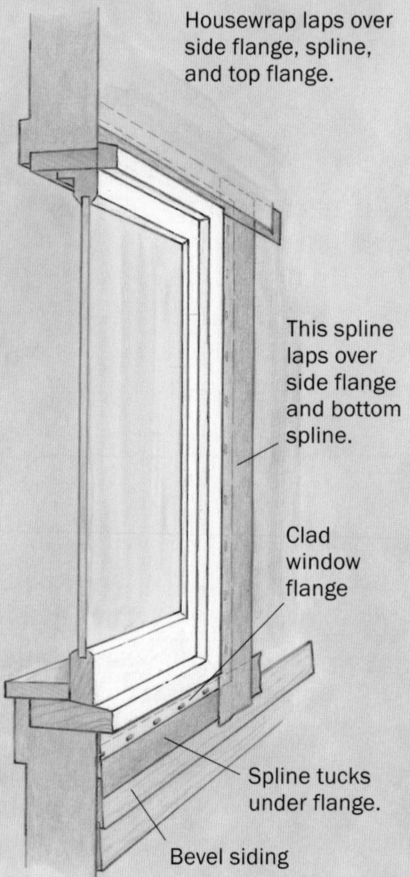

Metal Head Flashing for Windows and Doors

FOLDED HEAD FLASHING

Prefabricated aluminum flashing can be cut and folded on site (shown below).

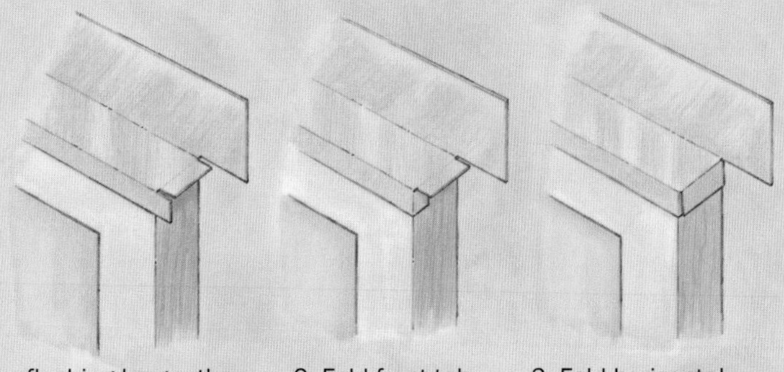

1. Trim flashing longer than head casing, notch back to casing, and fold in flashing.

2. Fold front tab back against side of casing.

3. Fold horizontal part of flashing down over side of casing.

SOLDERED HEAD FLASHING

The best head flashing is soldered at the end so that the end profile matches the side profile. Upper section of flashing extends past head casing.

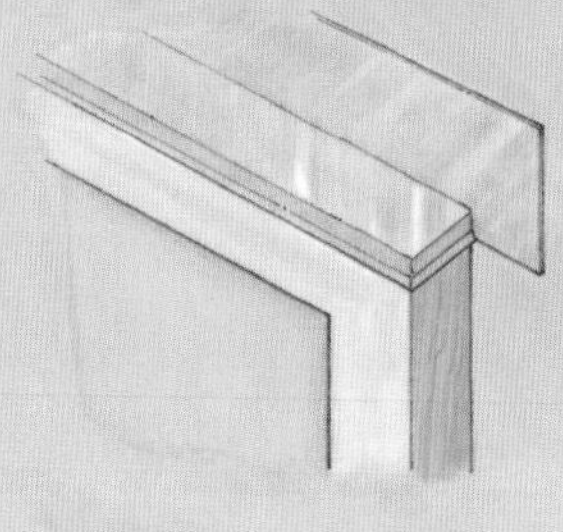

When siding reaches up as far as the bottom spline, the spline laps over the top edge of the last siding course before the next course goes on. That feeds any seepage quickly to the outside rather than let it creep behind the siding.

Doors are flashed the same way as windows except that the bottom spline should be replaced with metal that folds in a few inches over the subfloor. Because door bottoms sit closer to the moist ground, they require the superior protection of metal. A pan with edges that turn up around the doorsill and down over the housewrap provides maximum protection from water infiltration. Such pans can be soldered, but prefabricated plastic pans are also available for purchase (Jamsill®, P. O. Box 485, Talent, OR 97540; 541-488-7470).

Round-Top Windows Can Be Tricky

Most round tops today come clad in vinyl or aluminum, thus eliminating the need for a separate head flashing. That doesn't make them foolproof, however. As rain strikes a regular square-top window, most of the water will immediately run off the head flashing and down over the window. In the case of a round top, however, the water would rather travel laterally down along the more steeply pitched arc of the window, directing a concentrated flow to the outside corners of the round top. This flow should be shunted to the exterior if possible.

The best solution would be to let the window's lower horizontal nailing flange lap over the siding. Because this design would look bad, the next best thing is to make sure that a hidden piece of felt or sheet metal tucks under the nailing flange and laps over the top edge of the first available piece of siding under the window. In the case of stucco or vertical siding, there's really no clean, fail-safe way to flash to the outside; a good bead of caulk between the wall and the head flashing then becomes the main line of defense against water infiltration.

When installing round-top and square-top windows that are mulled together (or combined) to make a Palladian-style window, make sure that the head flashings have been properly sealed at the point where the arc meets the level head casing. That means caulking for aluminum-clad windows, or a solvent weld for vinyl clads. Leaks here are common.

In renovation work it is sometimes necessary to custom-flash an existing round-top window. To make custom flashings for round-top windows, I first bend an ordinary, straight Z-flashing on my brake. I then snip the flange that will eventually turn up under the siding, cutting at regular intervals depending on the tightness of the radius.

Compensate for the curves. Crimps in the bottom flange and cuts in the top flange allow the flashing to follow the bend of the arch-top window casing. Small squares of flashing are either caulked or soldered to fill the V-shaped spaces left by the cuts.

Filling the V-Notches in Curved Flashing

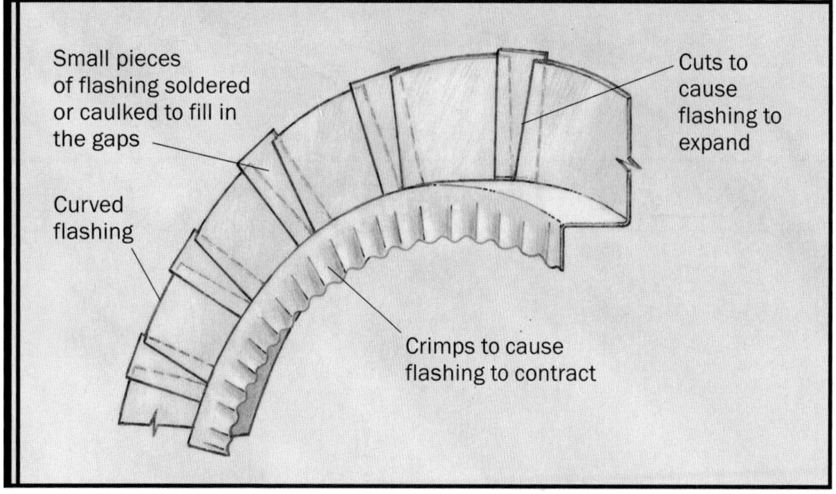

The flange that will turn down over the window gets crimped with a hand crimper. These steps effectively stretch the upturned flange and contract the downturned flange, causing it to bend (see the photo on p. 11).

During bending, the cuts in the upturned flange open to become V-notches. To fill the notches, I cut small squares of flashing and slip one into each notch, sort of like slipping cards into a poker hand. If I'm using copper, I can solder the "cards" in place. For aluminum it's the caulk gun (see the drawing on p. 11).

The crimped, downturned flange has a piecrust texture, which is generally unnoticeable when the window sits high in a gable. In cases where the crimping looks objectionable, the crenellations can be leveled off with solder or auto-body filler.

Step Flashing Keeps Water Running Down and Out

Where a sloped roof meets a sidewall, step flashing is used. Step flashings are small, rectangular pieces that are bent down the middle into right angles. Each course of roof shingles gets one step flashing (see the drawing below). In effect, the step flashing is just a flexible metal shingle that turns up under the siding.

Dormer Flashing

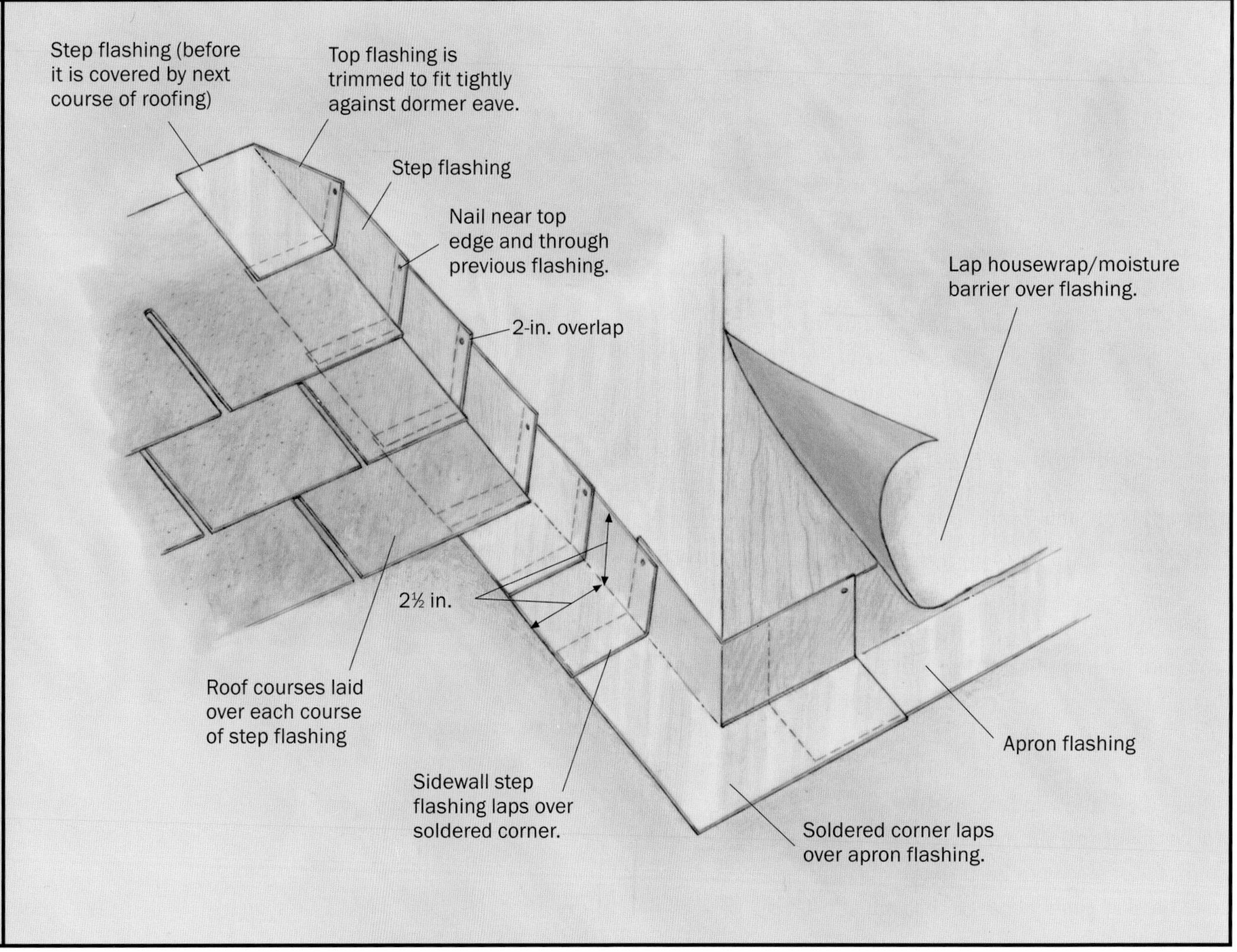

An Alternative to Soldered Corner Dormer Flashing

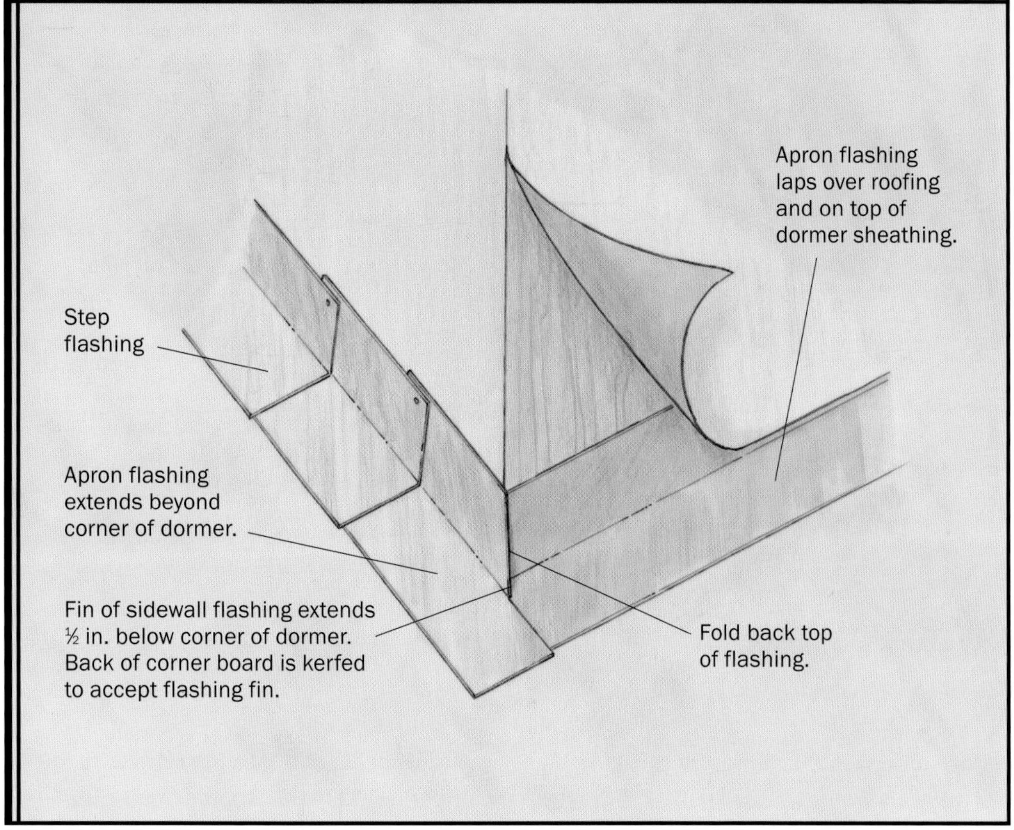
Apron flashing laps over roofing and on top of dormer sheathing.
Step flashing
Apron flashing extends beyond corner of dormer.
Fin of sidewall flashing extends ½ in. below corner of dormer. Back of corner board is kerfed to accept flashing fin.
Fold back top of flashing.

A typical step flashing measures 5 in. by 7 in. The 5-in. dimension gets folded in half so that 2½ in. turns up under the siding, and the other 2½ in. extends under the adjoining shingle. The 7-in. dimension runs downhill, stopping just above the butt edge of a roof shingle. Because the standard exposure for asphalt shingles is 5 in., 7-in. step flashings will have a 2-in. overlap. One nail is driven through both step flashings where they overlap on the sidewall. Siding then comes down over the upturned sides of the step flashing. I hold the siding ½ in. above the roof to keep the edges of the siding dry.

Flashing a Dormer Isn't Difficult

To flash a dormer, I first shingle as far as the front of the dormer. Then I apply the apron flashing (see the drawing above). This piece laps about 4 in. over the roofing and up about 4 in. onto the front of the dormer. When I'm working with copper, I can solder a separate piece to the apron for a positive wraparound seal at the corner. When I'm working with aluminum, I let the lowest piece of step flashing extend beyond the corner by ½ in. or so, as a sort of fin. Later, I kerf the back to accommodate the fin. After I install the apron, the sidewalls of the dormer are step-flashed. The top step flashing gets angle-trimmed to fit tightly under the dormer roof overhang.

__Scott McBride__ is a contributing editor to Fine Homebuilding *magazine. His book* Build Like a Pro: Windows and Doors *is available from The Taunton Press. McBride has been a building contractor since 1974.*

How to Avoid Common Flashing Errors

■ BY JAMES R. LARSON

To understand flashing, try thinking about a house as essentially a tent with architectural embellishments. You'll know what I mean if you've ever been camping in the rain and inadvertently zipped up the door flap so that the bottom edge turned inward. The big puddle inside the tent, the one that soaked your last dry shirt, was a flashing problem not unlike those you can build into a house if you're not careful.

Flashing is one of the smallest parts of a building, so it sometimes doesn't get the respect it deserves. But it sure will exact its revenge when it is overlooked or improperly integrated with other building elements. I see a lot of that. For the past seven years, I've worked as an architectural consultant, and I've poked into many moisture-damaged houses whose troubles started with improperly applied flashing.

Water is a wily enemy. Driven by wind or sucked into crevices by negative air pressure inside the house, it can flow uphill between construction layers. Once inside, water will subject building materials to alternating spells of soaking and drying until first the paint and finally the wood just give up.

Flashing to me is more than metal. In a general sense, flashing refers to a configuration of materials that are arranged to direct water to the exterior. The materials could be metal, asphalt-impregnated building paper (felt) or, more recently, adhesive-backed bituminous tape. Together, these materials work to ensure that corners, openings and edges will exclude water. The wall tape (see the photo on p. 16) is an important ingredient. It's a modified bituminous material on a backer of cross-laminated polyethylene with an aggressive adhesive. It's sold by several companies (see "Sources" on p. 21). Building felt, housewrap, and bituminous tape are part of what the Uniform Building Code calls a "weather-resistive barrier," the final barrier to water penetration.

Flashing should be applied so that water flows to the exterior rather than being trapped behind one of the building's construction layers. If the correct installation of flashing could be boiled down to a single

Window Nailing Flanges Left Unsealed

PROBLEM
Siding applied directly over nailing flanges on aluminum- or vinyl-clad windows is a practice that invites leaks. Flanges should be sealed with strips of building felt or adhesive-backed bituminous tape before the siding goes up.

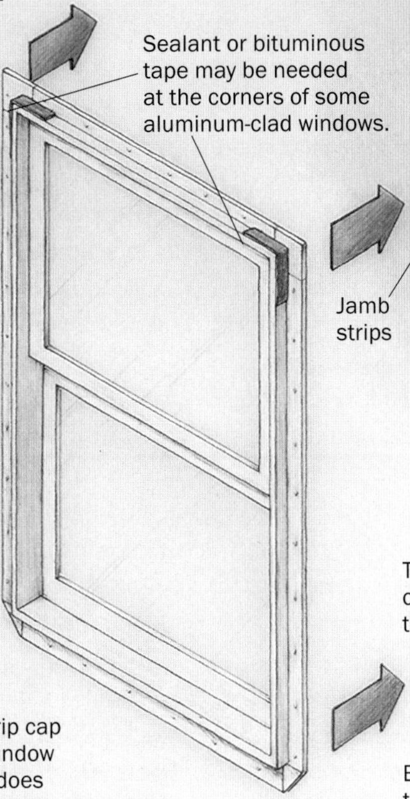

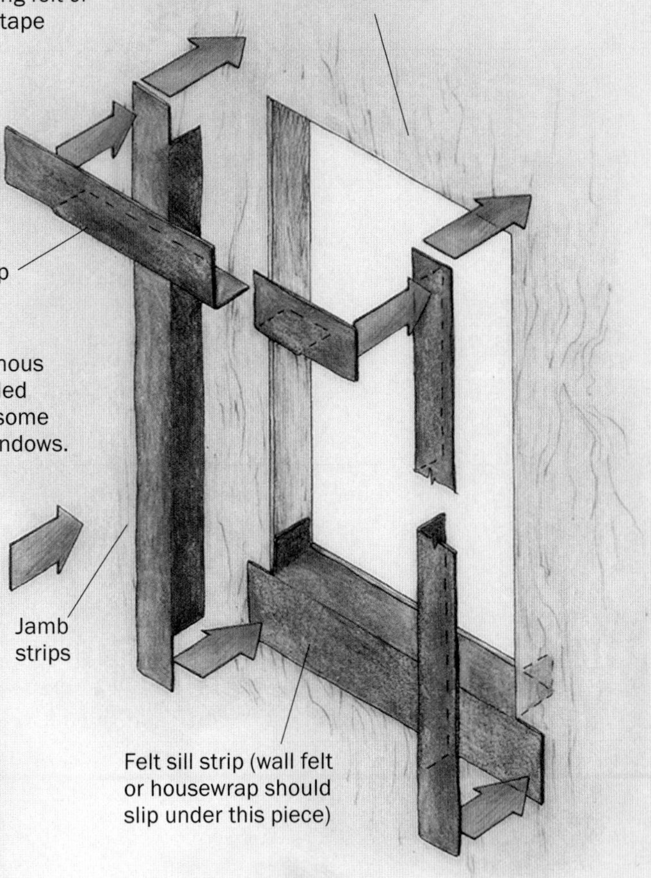

SOLUTION: WHEN WINDOWS GO IN BEFORE HOUSEWRAP
If windows are placed in their openings before any wall felt or housewrap has been applied to the sheathing, felt prep strips should be applied first. Make sure that the corners of aluminum-clad windows are well sealed.

Note: Add a metal drip cap at the head of the window if the manufacturer does not warrant the flanges as weatherproof.

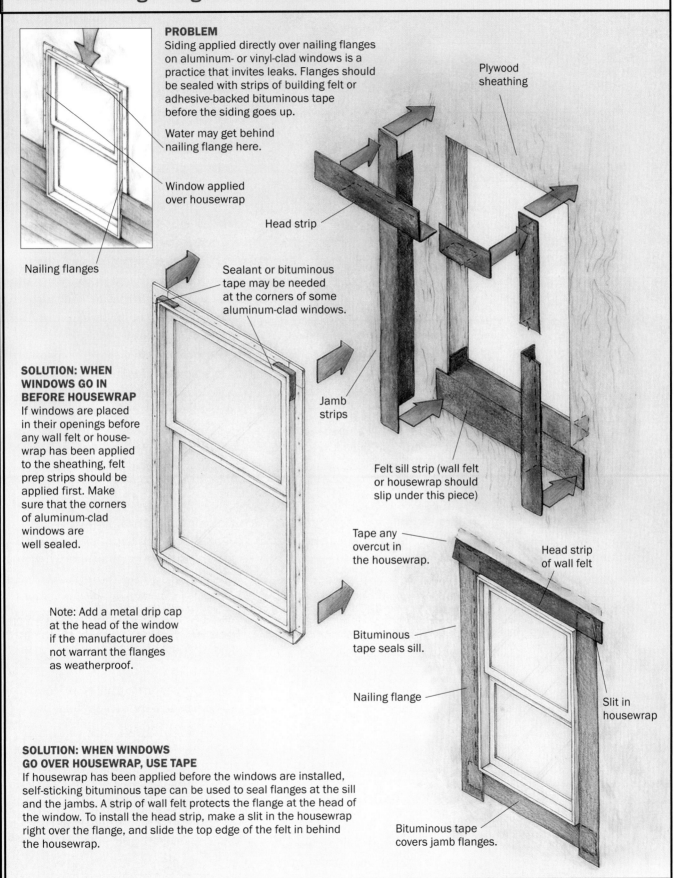

SOLUTION: WHEN WINDOWS GO OVER HOUSEWRAP, USE TAPE
If housewrap has been applied before the windows are installed, self-sticking bituminous tape can be used to seal flanges at the sill and the jambs. A strip of wall felt protects the flange at the head of the window. To install the head strip, make a slit in the housewrap right over the flange, and slide the top edge of the felt in behind the housewrap.

It's sticky and waterproof. Self-adhering bituminous wall tape has a variety of flashing applications, handling everything from window flanges to deck ledger boards.

If the correct installation of flashing could be boiled down to a single phrase, it would be this: Don't leave any edges looking uphill.

phrase, it would be this: Don't leave any edges looking uphill. This underlying principle is about as basic as placing the shower curtain on the inside of the tub before you turn on the water. In my dealings with problem-flashing installations, I've found four situations where mistakes commonly occur: around vinyl-clad and aluminum-clad windows manufactured with nailing flanges, at the intersection of a roof and a sidewall (a dormer, for example), where an outside deck intersects the sidewall of a house, and, finally, at the edge of a roof. Correctly flashing these spots won't take much more time than doing it wrong, and the effort will save you lots of heartache.

Window Nailing Flanges Should Be Sealed Before They Are Covered Up

Back in the '50s, wooden windows arrived on the job site with brick molding already attached. After strips of building felt were put in place around the window opening, the window could be hoisted into place and nailed through the brick molding. A metal flashing cap lapped behind the final wall felt protected the head of the window. Aluminum-clad and vinyl-clad windows now arrive with nailing flanges around the perimeter. These flanges must still be flashed and sealed before siding goes on to prevent water and air infiltration. My rule of thumb is that if you can still see the nailing flanges at the jambs and at the head, you're not yet ready to install the siding (see the top left drawing on p. 15).

There are two possible scenarios here: first, when windows are installed directly over the sheathing; and second, when windows are placed after the building is covered with building felt or housewrap.

When windows are installed over sheathing, strips of felt should go on all four sides of the opening just as in the 1950s version. The flanged window is nailed through these prep strips; then the felt or housewrap goes on. The sill strip should lap over the housewrap at the bottom. If the window manufacturer does not warrant the flange as weatherproof flashing, a metal drip cap should be tucked in behind the housewrap at the head. No tape or sealants are required.

Lately, I've been seeing a more high-speed variation of this detail where only the sill prep strip has been used. The flanges are then sealed to the sheathing with either a bead of sealant between the flange and sheathing or sealed with strips of bituminous wall tape before being covered with the weather barrier. This variation trades in the principle of redundancy for a faith in chemical seals. But I have to admit that it seems to work effectively.

More commonly, houses are sheathed and covered in housewrap before the windows are even delivered. The windows get nailed on right over the housewrap, and before you know it, the siding has been installed. You have been left with the nailing flange at the head of the window on top of the housewrap. The top of the flange is now ready and willing to channel into the wall any water that passes its way.

One answer, as shown in the bottom drawing on p. 15, is to make a slit in the housewrap just above the head flange after the window unit has been nailed in place. The slit should extend 6 in. beyond each side of the window frame. A 6-in.-high strip of felt is then slipped into the slit and brought down over the outside of the head flange. Strips of the bituminous tape 6 in. wide should be used to seal the sill and jamb flanges to the housewrap. (Don't use red vapor-barrier tape for this application—it was designed primarily for interior use, and the adhesive won't take too much water.) Applying a felt strip at the bottom of the opening before setting the window in place, just as you would if you were installing the window on bare sheathing, is a way to improve this installation method.

The most high-speed technique of all is simply to nail on the window flanges over the housewrap and then to seal the edges with bituminous tape, starting with the sill, then jambs, then head. This technique epitomizes an almost-complete departure from the traditional reliance on redundancy to a faith in new materials and adhesives. Unfortunately, at this point we do not have a 50-year history of performance of these materials to tell us whether our faith is well placed.

Do you need a metal flashing cap, too? Probably not, but don't forget to look at window corners carefully. Some aluminum-clad windows may show small gaps at the corners where the miter cuts in the metal cladding have separated slightly during shipping. Gaps, no matter how small, can cause serious water leaks. I'd add a patch of the bituminous wall tape if there's any hint of an open joint. Vinyl-clad windows generally don't have this problem because they are made with heat-welded corners.

Another vulnerable point of aluminum-clad windows is the corners where the side flanges meet the top and bottom flanges. Manufacturers typically provide some sort of corner clip for these spots, but even with the clips in place, there is usually a pesky little hole about 1⁄16 in. across that doesn't get covered. This hole needs to be covered with either the bituminous tape or with a dot of exterior-rated sealant.

The most high-speed technique of all is simply to nail on the window flanges over the housewrap and then to seal the edges with bituminous tape, starting with the sill, then jambs, then head.

When Roof Felt Does Not Lap Up an Intersecting Wall, Trouble May Follow

A sloped residential roof often butts into a vertical surface, such as a dormer or a chimney. At the juncture of the roof sheathing and the vertical surface, there will be a crack—a perfect entry for water if this seam isn't sealed carefully (see the left drawing on p. 18). Metal step flashing isn't enough. Wind-driven rain or ponded water blocked by a snowbank or ice dam can creep up underneath metal flashing and get into the building.

Anyone who has worked with roofing underlayment knows how unwieldy long pieces of this material can be. It's no easy feat to take a 24-ft. piece of felt and get it on straight—it's even harder to do when you have to lap it up the bottom wall of a dormer by 4 in. before attaching it. But as the drawing shows, that's exactly what should be done (see the center and right drawings on p. 18). The roof underlayment should always extend up an intersecting wall by this amount.

The metal flashing is next in sequence (step flashing along with shingles on the sidewalls and a continuous apron flashing along the bottom edge of a wall). This should be followed by building paper or housewrap on the sidewall, which should lap down over the top edge of the roof underlayment to complete the seal.

Roof Underlayment Ends at Intersecting Wall

PROBLEM
Conventional metal step flashing alone does not offer enough protection from water infiltration. Wind-driven rain or ponded water on an ice-covered roof can back up beyond the top edge of flashing and get inside the building.

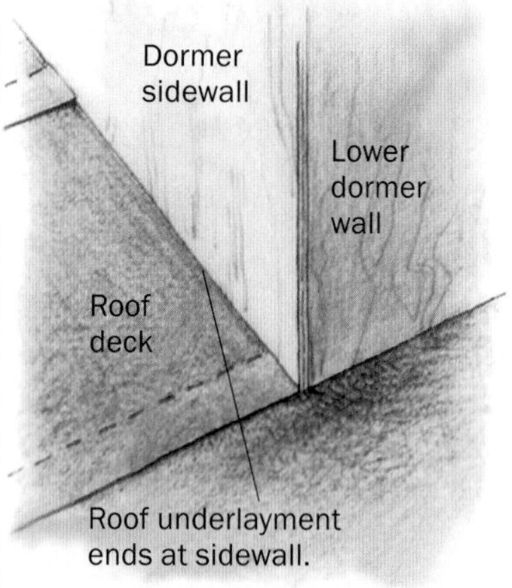

SOLUTION: LAP ROOF UNDERLAYMENT UP INTERSECTING WALLS
A better seal is provided when roof underlayment is lapped 4 in. up any intersecting vertical walls. Follow that with conventional metal step or apron flashing with the shingles.

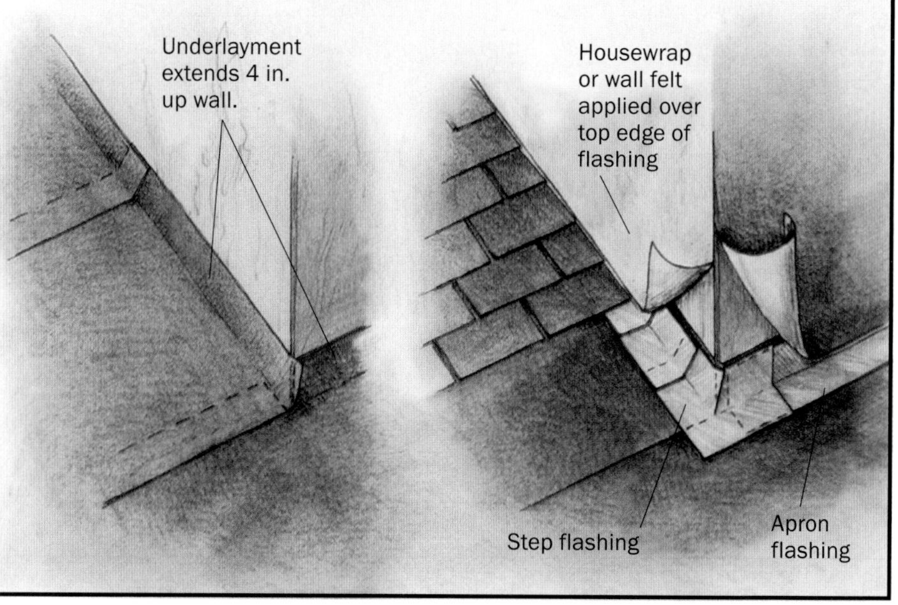

I recently worked on a house that leaked along the bottom edge of a shed dormer, not regularly but enough to puzzle the owner and the roofer about the source of the water. A hard wind-driven rain could be forced up the wall just enough to leak inside the building. We were able to loosen the metal apron flashing and the metal step flashing and shingles. We installed a strip of self-adhering bituminous tape from the vertical surface down onto the roof felt and replaced the metal flashing and shingles. Problem solved.

Lapping the roof underlayment up the adjoining wall is an approach that has the virtue of common sense. But trying to find an industry publication depicting it turns out to be a bit of a research project. You will, however, find a straightforward description of how to join a roof slope to a wall in the last place you might think of looking: right on the brown-paper wrapper that roof shingles come in. The one I'm looking at now is one I picked up out of my own yard after the roofers had finished work. It's from GAF® Materials Corporation and suggests that the roof underlayment extend 4 in. above the roof/wall intersection. The instructions also advise adding asphalt plastic cement behind the turned-up felt. Why this information doesn't get transferred from wrapper to roof more often is a mystery.

Insufficient Flashing Along the Edges of a Deck Usually Leads to Rot

A project I recently looked into involved a typical condition, rot where a deck had been added along an exterior wall of a house. The deck was solid. But the wall had been leaking since the deck was installed. By the time I saw it, the sheathing had the consistency of soggy Shredded Wheat. This house had

No Wall Felt, Inadequate Flashing at Deck

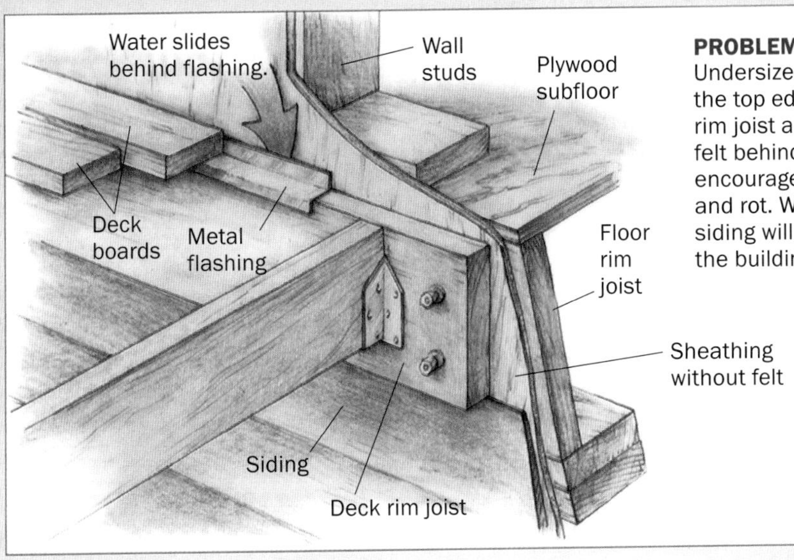

PROBLEM
Undersized flashing along the top edge of a deck's rim joist and a lack of wall felt behind it are likely to encourage water infiltration and rot. Water driven past siding will end up inside the building.

SOLUTION: ADD BITUMINOUS TAPE AND WALL FELT TO PREVENT LEAK.
Adhesive-backed bituminous tape at the sill is the start of a leakproof deck installation. The tape should lap down over the outside edge of the deck rim joist and up the wall 4 in. Housewrap or wall felt should cover the top edge of the tape.

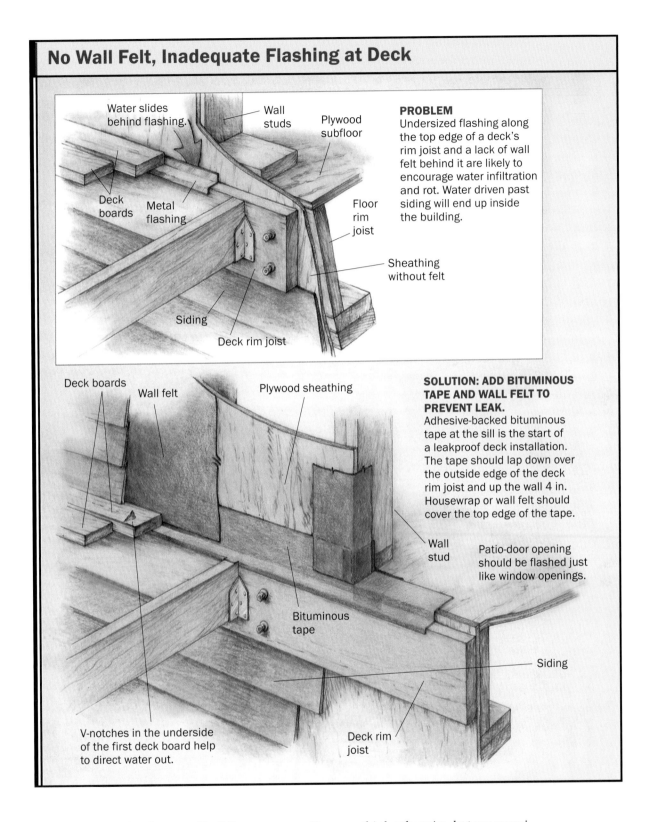

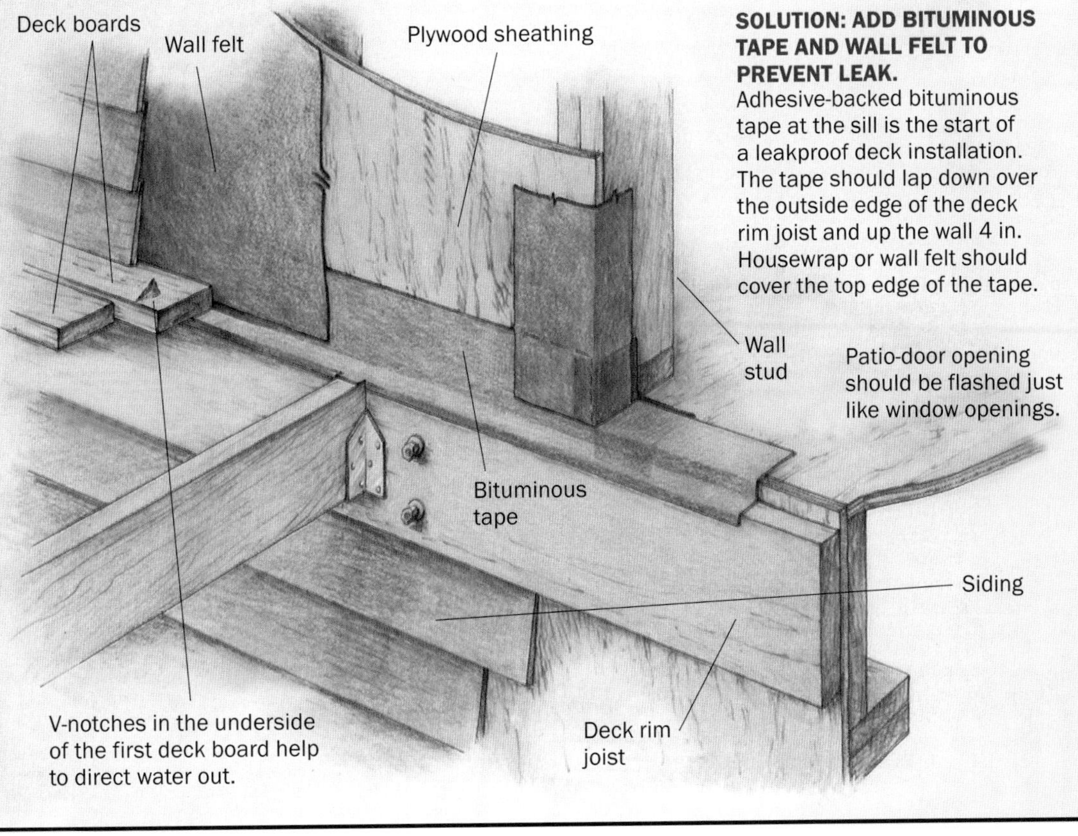

three problems: the absence of building paper or housewrap over the sheathing, flashing at the edge of the deck that was too narrow, and a patio door that had been installed so that it allowed water leaks around the edges. The top drawing above shows details.

You may think otherwise, but my experience has been that a small amount of water often, if not always, gets past siding. It may be driven by the wind or sucked in by capillary action. In the house I looked at, some water probably was making its way behind the siding in this way, and then getting behind the flashing at the bottom of the wall. The builder originally had caulked between the bottom siding board and the deck, but some water also may have been driven in here. The flashing did not extend far enough up the wall to prevent water from getting into the framing. Any water coming from higher up the wall encountered no weather barrier to direct it out over the metal flashing.

To prevent these leaks, start with a 6-in. wide strip of bituminous wall tape as the base flashing, as shown in the drawing (see the bottom drawing on p. 19). The wall tape is a much better choice than metal flashing: It's self-sealing around nails and at lap joints and splices, it's easily formed, and it won't corrode or dissolve. (The original purpose of this material, in a wider form, was to make pond liners.) It is applied so that it overhangs the front edge of the deck rim joist slightly and goes up the sheathing by about 4 in. The first deck board is notched on the bottom at regular intervals to allow any water that gets in there a way to get out. You could just cut saw kerfs every 16 in. or so on the bottom of the board. But a better way is to set your circular saw to a 45-degree angle and make two cuts to produce V-shaped notches. They won't be as easily blocked by debris that happens to get in there.

Once the base flashing is applied, the building paper on the outside of the building should be lapped over the top edge to complete the seal. Patio doors should be flashed like windows, with sill, jamb strips and head strips applied so that they direct water to the outside.

Standard Metal Drip Edge Won't Always Prevent Leaks

Metal roof edge, also called drip edge, installed along the rakes and eaves of a roof establishes a smooth, clean surface and a straight line where roofing felt and shingles

Metal Roof Edge Isn't Wide Enough

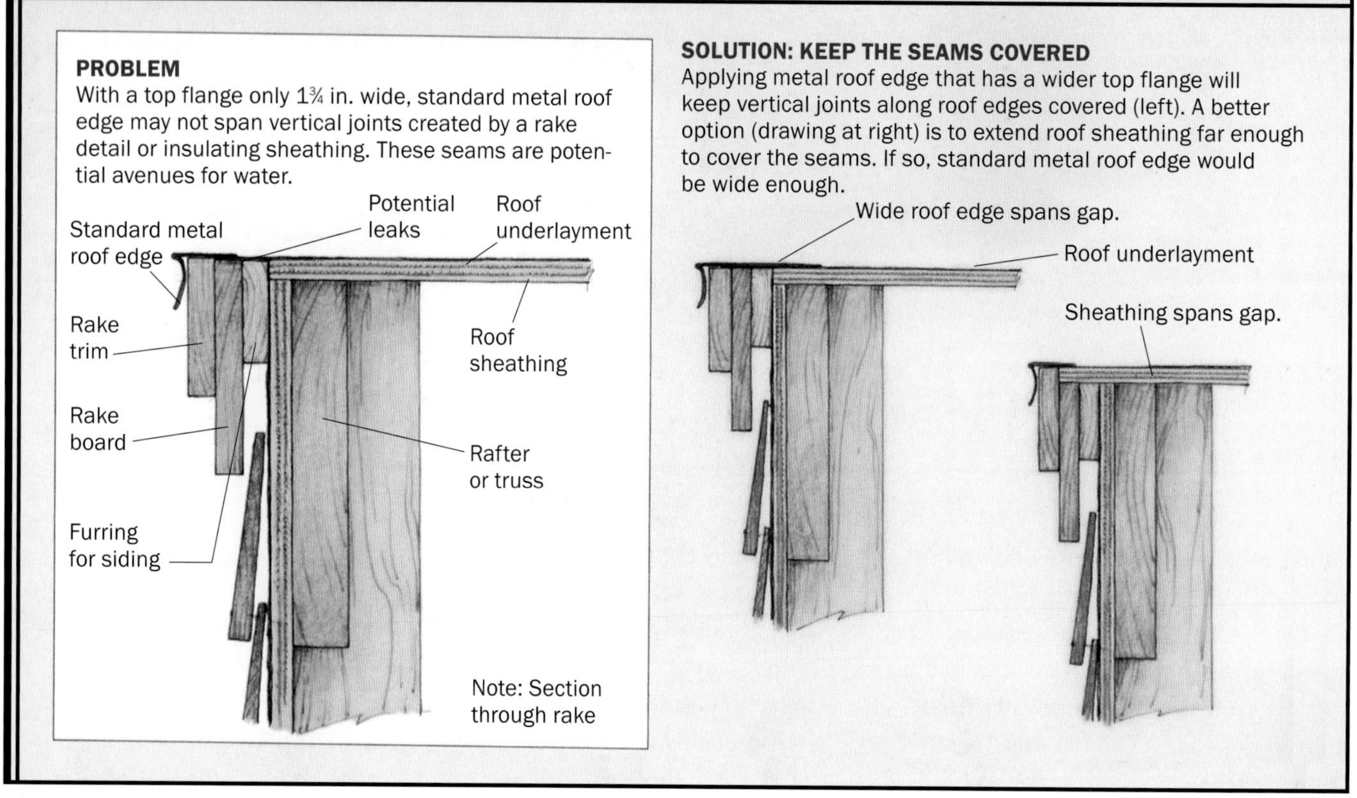
PROBLEM
With a top flange only 1¾ in. wide, standard metal roof edge may not span vertical joints created by a rake detail or insulating sheathing. These seams are potential avenues for water.
Standard metal roof edge
Potential leaks
Roof underlayment
Rake trim
Roof sheathing
Rake board
Rafter or truss
Furring for siding
Note: Section through rake
SOLUTION: KEEP THE SEAMS COVERED
Applying metal roof edge that has a wider top flange will keep vertical joints along roof edges covered (left). A better option (drawing at right) is to extend roof sheathing far enough to cover the seams. If so, standard metal roof edge would be wide enough.
Wide roof edge spans gap.
Roof underlayment
Sheathing spans gap.

can end. It also covers over splice joints in the fascia board or the gap created at the junction of the fascia board and the roof sheathing. In typical metal roof edge, the top flange is about 1¾ in. wide. With some types of insulating sheathing or trim details, however, a gap can be created behind the fascia that is too wide to be covered by standard metal roof edge (see the left drawing on the facing page).

Sometimes a dip can develop in the roof felt and shingles near the roof edges (see the left drawing below). This depression can encourage ponding of water driven into this area, and it increases the chance that water will work its way behind trim or into walls. If you're tempted to overlook this little glitch, don't. I've been called out to problem houses where that was the case. It takes about three years for the paint to peel off and four years for staining to start. Degradation of the wood follows shortly thereafter.

If the framers have left too wide a gap from the edge of the roof deck to the fascia, your best bet is to use an extra-wide metal roof edge. It will completely cover any gaps. In these situations, I recommend roof edge with a 4½-in. flange that extends onto the roof deck.

In many cases, the builder wants the shingles installed before the siding or fascia is in place. It is well worth the extra few minutes it takes to figure out the thickness of any insulation and fascia so that the metal roof edge can be properly placed to receive these materials.

James R. Larson became an architect over 30 years ago. For the past 13 years, he has had a solo practice as a consultant to builders, homeowners, and other architects. He lives in St. Paul, Minnesota.

Sources

Self-adhering bituminous wall tape is available from these three companies.

BarTech International
3441 S. Willow Ave.
Fresno, CA 93725
(800) 341-9917

Grace Construction Products
(800) 558-7066
www.na.graceconstruction.com

Protecto Wrap® Co.
2255 S. Delaware St.
Denver, CO 80223
(800) 759-9727

A Sag May Mean Ponded Water

PROBLEM
Blocking nailed to a rake edge or an eave to support a cornice may create a dip in the roof underlayment. Standard metal roof edge doesn't cover the gap.

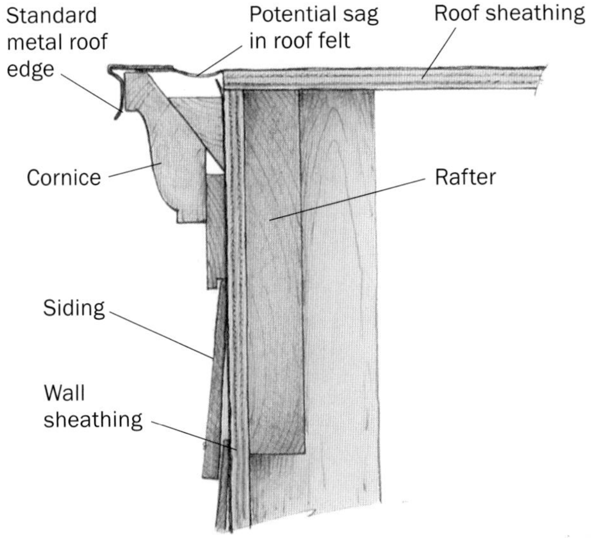

SOLUTION: WIDE ROOF EDGE SPANS GAP
Metal roof edge with a 4½-in. flange will span any gap between a cornice detail and the roof sheathing, preventing sag in roof underlayment and shingles and cutting off a possible source of water leaks.

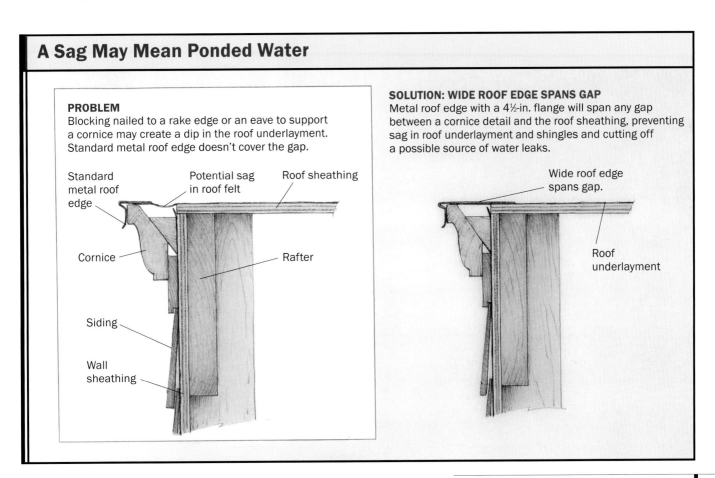

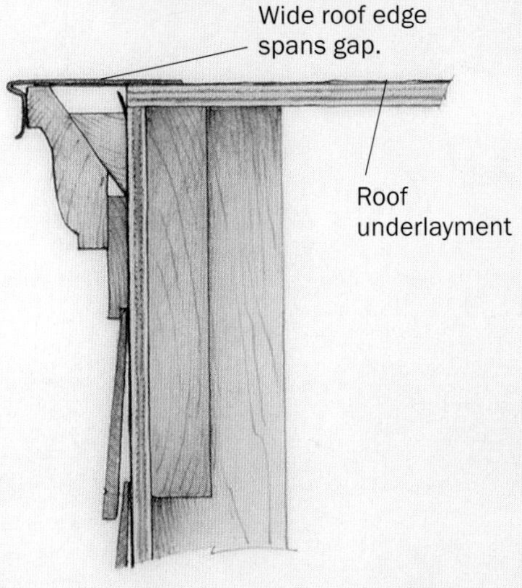

All About Rain Gutters

■ BY ANDY ENGEL

You've probably cursed your gutters more than blessed them. Purposeful yet troublesome, gutters are conspicuous as architectural afterthoughts on the otherwise carefully designed facades of many houses. Gutters clog with composting leaves, and in my neighborhood, they often sprout thickets of maple seedlings. In the north country, gutters that are half-torn from the house by sliding snow predict spring's arrival more accurately than does any groundhog.

Despite their shortcomings, gutters are essential to the longevity of most homes. Without gutters and downspouts leading rainwater away from the house, foundations become feedlots for mold that can sicken your family and rot your house. Wet foundations lead to peeling paint and even to damp attics.

Unless the soil that is surrounding your house is free-draining gravel that never saturates, or unless you live where rain is only a Christmas-tree decoration, your house needs gutters. Here's how to make the best of them.

Sizing Gutter Systems

Gutter-system design takes into account likely rainfall intensity, roof size, gutter volume, and downspout size and frequency.

1. Calculate your roof's watershed area

A roof's watershed area isn't obvious. Maximum rainfall is likely wind driven, so steep roofs may collect more water than flat roofs. To figure a roof's watershed area, multiply its surface area by the appropriate factor from the table at right.

12-in-12	1.3
9-in-12 to 11-in-12	1.2
6-in-12 to 8-in-12	1.1
4-in-12 to 5-in-12	1.05
Flat to 3-in-12	1

2. Find the maximum likely rainfall intensity

Residential gutters are often planned to handle the most intense five-minute burst of rain, measured in inches per hour, that's likely to occur in a ten-year period. Find yours from the map.

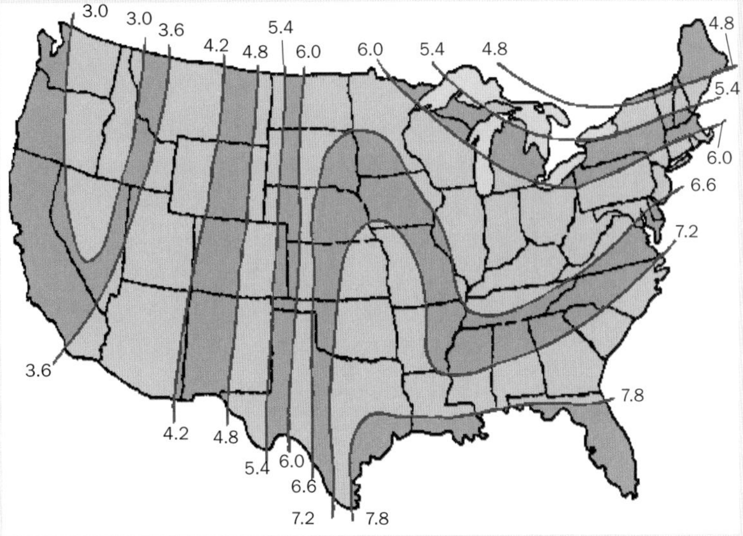

3. Determine the gutter needed to drain your watershed

Divide your favored gutter's 1-in.-per-hour watershed (see the table at right) by the five-minute rainfall intensity (from 2, above). This determines the maximum watershed level gutters can serve between downspouts. Pitch your gutters by ⅛ in. per ft., and you can multiply the watershed by 1.4.

Each square inch of downspout cross section can drain 100 sq. ft. of watershed. So a 2-in. by 3-in. spout drains up to 600 sq. ft., and a 3-in. by 4-in. spout drains 1,200 sq. ft.

Going from one downspout to two doubles the watershed that a section of gutter can drain.

Watershed gutters drain at 1 in. rain per hr.

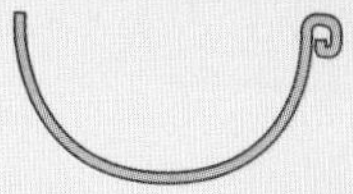

5-in. half-round	2,500 sq. ft.
6-in. half-round	3,840 sq. ft.

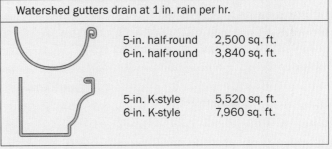

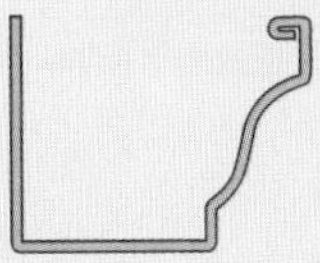

5-in. K-style	5,520 sq. ft.
6-in. K-style	7,960 sq. ft.

Sample house

An 8-in-12 pitch shed roof in Washington, D.C., is 40 ft. wide, and its rafter length is 20 ft. The roof's area is 800 sq. ft. The pitch factor for an 8-in-12-pitch roof is 1.1; when multiplied by 800 sq. ft., that gives a watershed of 880 sq. ft. The theoretical 5,520-sq. ft. watershed drained by a 5-in. K-style gutter, divided by Washington's 6.6-in.-per-hr. rainfall intensity, shows a maximum watershed of 836 sq. ft. Close, but to be safe, the builder should either pitch the gutter, use a larger gutter, or add another downspout.

Aluminum Gutters Are the Most Popular

The K-style is frequently seamless, formed on site with a mobile machine. Installed for around $3* per ft., these gutters are the standard. They're also available in 10-ft. or 20-ft. lengths, K-style, or half-round, and they can be riveted together and sealed with silicone caulk.

Aluminum-gutter stock comes in several weights, the most common being 0.028 in. and 0.032 in. It usually doesn't cost much more to upgrade to the heavier 0.032 in., and the gutter will be harder to dent and will hold up better to heavy snow and ice.

Aluminum gutter stock comes prepainted, most commonly in white. Check with your supplier for other colors. Aluminum gutters last. My parents' gutters are 30 years old, dented and dull-colored, but water still flows through them as if they were new. Aluminum can be repainted; however, replacing them might not cost much more.

There's More to Choose From than Seamless Aluminum

One of the first choices you'll have to make is what material and profile your gutters should be. The most common materials are aluminum, copper, galvanized steel, and plastic. The Architectural Sheet Metal Manual (SMACNA/Sheet Metal and Air Conditioning Contractors National Association Inc.℠; 703-803-2980) lists over a dozen standard profiles. (At $176, this book is pricey, but it shows every flashing and sheet-metal detail imaginable.)

The most common gutter in use nationally is 5-in. aluminum K-style (see the photo at left). K-style gutters are called by that name simply because the profile's place in SMACNA's alphabetical hierarchy is the 11th letter of our alphabet. The seamless-gutter contractors that I know produce miles of this rectangular-back, ogee-front gutter every year.

Seamless gutters are the least likely to leak. Specialized truck-mounted or trailer-mounted forming machines pull flat metal stock from a coil and shape gutters of the desired profile on site. One-piece lengths as long as the stock on the coil are possible, but thermal expansion and contraction limit the practical length to about 50 ft. These forming machines are costly, and contractors are likely to be able to produce only 5-in. and maybe 6-in. K-style.

Of course, lumberyards and specialty wholesalers sell these and other profiles already formed, but their lengths are usually limited to 10-ft. or 20-ft. sections that must be joined on site.

The problem with joining gutter sections is that joints are leak-prone. On metals that can be soldered, copper or galvanized steel, the sections must overlap by 1 in. and be riveted on 2-in. centers before soldering.

Long-Lasting Copper Is Pricey

A commodity metal, copper sells by the pound, and its price fluctuates with the futures market. As of this writing, 5-in., 16-oz. half-round gutters cost $4.20 per ft. locally. Copper's thickness is designated by weight. Sixteen-ounce sheet copper, standard for gutters, weighs 16 oz. per sq. ft.

Copper makes durable but easily dented gutters. Examples have lasted more than 50 years in corrosive seacoast environments. Copper's new-penny luster develops a green patina within a few years, and water dripping from copper can stain lower surfaces green. Runoff from cedar-shingle roofs is said to corrode copper fairly quickly.

Copper gutter can be formed seamlessly with the same equipment used to form aluminum gutters. It's also available in 10-ft. and 20-ft. lengths preformed as half-round and K-style. Being very malleable, copper lends itself to custom shapes and linings for concealed gutters.

Steel Gutters Are Hard to Dent

I worked on a 75-year-old house whose galvanized gutters leaked only at the lower-downspout elbows. There, swiftly moving water had worn away the protective zinc, and rust holes developed. The outside of these gutters hid under a paint layer thick as a tortilla chip, but even the unpainted gutter bottom had only begun to rust.

Exposed steel rusts. Steel can be galvanized with zinc or it can be protected with Galvalume®, a proprietary coating of zinc and aluminum, and with terne, a tin and lead alloy. Galvanized is the cheapest, about 70¢ per ft.

Steel gutters are available painted with the same slick, durable finishes found on commercial roofs. Steel is strong, and gutters made of it stand up well to the weight of ice and snow. Although K-style and half-round profiles are common, others are used on commercial projects and can be adapted to residential use.

No Corrosion With Plastic Gutters

The newcomer to gutter materials is plastic, generally similar to the material vinyl windows are made of. Although many of us shy away at the V-word, vinyl gutters, particularly the half-rounds, look pretty much like other gutters. They don't corrode, and as long as they include ultraviolet inhibitors, vinyl gutters should last a long time.

I feared plastic would become brittle when cold, so I left a sample in the freezer overnight. The next morning, it was still flexible, and it survived unscathed when I whacked it hard on my desk.

Supplied in 8-ft. or 10-ft. lengths, vinyl gutters seal together either with integral neoprene gaskets or with cement similar to that used for plastic pipe. In either case, the manufacturers include some way to accommodate expansion. Vinyl expands and contracts more than any metal. Its price isn't bad, about $2.25 per ft.

Aluminum and painted steel can't be soldered. Their 1-in. lap must be sealed with paint-compatible gutter sealant or high-grade silicone caulk and riveted on 1-in. centers.

Custom gutters can also be bent on site with a sheet-metal brake, but when they're made that way, the section lengths are limited by the width of the brake, typically 10 ft. or less. If you're custom-bending gutters, the more longitudinal breaks you add to the profile, the stiffer the gutter will be. Make sure the front of the gutter is at least 1 in. lower than the back. Then, if the gutter fills up, it will overflow away from the house.

Half-Round Gutters Drain Better

Joining issues aside, profile choice is largely a matter of taste. Half-round gutters, for example, seem to go with several turn-of-the-century architectural styles. Square or angular gutters can work well on contemporary houses. A crown molding run below K-style gutters blends them into a fancy colonial cornice.

Half-round gutters may drain more completely than flat-bottom gutters. This fact makes sense when you think about how large an area exists at each gutter type's bottom, where water can linger. Why should you care about a little stagnant water in your gutters? In a word, mosquitoes. The solution is simple, though. Keep flat-bottom gutters free-flowing by substantially pitching them and cleaning them scrupulously.

Extending the roof's drip edge into the gutter is a good practice. It ensures runoff that may be drawn along the shingle's bottom by capillary action won't drip between the house and the gutter.

The drip edge stocked by my local lumberyard extends only about 1 in. below the roof, not far enough to reach into a more than minimally pitched gutter. Simply lapping a piece of flashing over the back of the gutter and under the drip edge makes up for this shortcoming. Gutters that have a straight back, such as K-style, lend themselves to this detailing. The brackets for half-round gutters usually push them far enough off the fascia to make running the drip edge into the gutter problematic.

K-style gutter is sometimes available with an integral flange at its back. The flange slides under the roofing material, much as the flashing detail I described for drip edge and gutter. Gutters of this design, though, must be installed parallel with the roof edge—that is, level, unless you've had the forethought to custom-cut the rafters so that the roof pitches sideways.

An Expansion Joint Keeps Gutters From Wrinkling

Gutters installed in extremely hot or cold weather, and those longer than 50 ft., may require some provision for expansion or contraction (see the sidebar below). A 50-ft. aluminum gutter installed on a 0°F winter day will grow ¾ in. in length on a 100°F summer day. A long gutter rigidly anchored on both ends, such as one that rounds two corners, is subject to expansion and contraction woes. Temperature-induced expansion or contraction can cause the gutter to wrinkle, or stress the seams, possibly to the point where the seal breaks and leaks begin.

The solution to this problem is an expansion joint in the run. Although this sounds complicated, it's not. The gutter is divided in two, and both ends are capped. The ends are separated by the expected maximum expansion, based on the length of the gutter and the temperature when it's installed. An expansion joint dams the gutter, so a downspout is needed for each section.

Accommodating Expansion

Steel: 0.0000067 in.
Copper: 0.0000094 in.
Aluminum: 0.0000129 in.

These numbers, coefficients of expansion, describe how much 1 in. of these materials expands or contracts with each degree of temperature change. For example, a 50-ft. steel gutter installed when it's 50°F will be 0.2 in. shorter at 0°F, and 0.2 in. longer at 100°F; 50 ft., or 600 in. x 0.0000067 in. x 50°F temperature change = 0.2 in.

An expansion joint (right) lets gutters move without stressing the seals. To accommodate movement, simply divide gutters more than 50 ft. long in two, cap the ends, and provide downspouts for each.

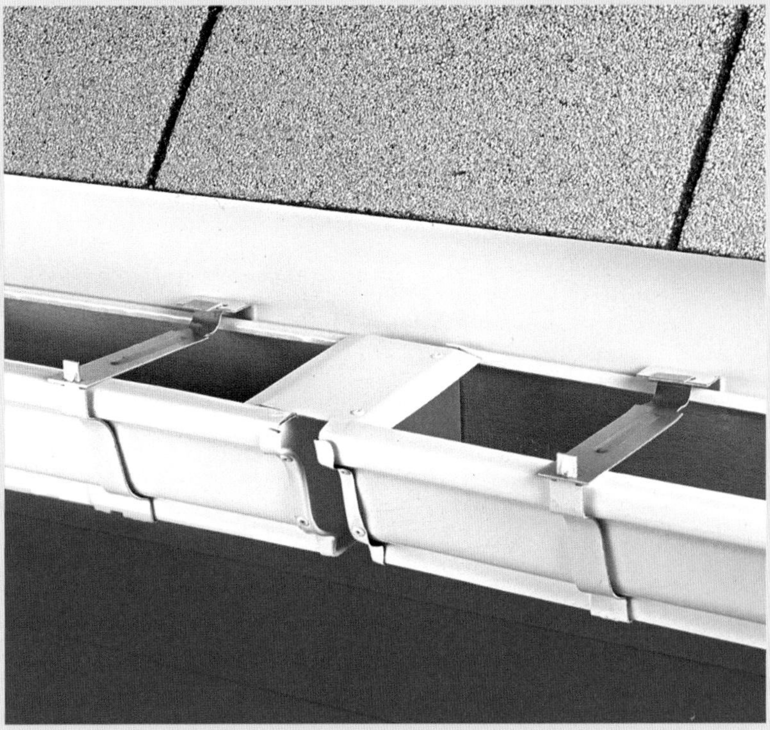

Narrow Downspouts Clog More Than Wide Downspouts

A few installation details can ease cleaning and minimize clogs. The simplest trick is to install wider gutters. It's easier to fit your hand inside a wide gutter than a narrow gutter. Also, wide gutters can handle more runoff and require more leaves to clog.

Downspout clogs can start when debris catches on the screws that hold spouts together. Blind rivets protrude less into the flow and are less likely to catch twigs and leaves. But if a clog develops, you have to drill out the rivet to disassemble the downspout for cleaning.

The best solution is probably to rivet the elbows to the downspout, and then to screw the assembly to the gutter so that it can be removed for cleaning. Use the shortest screws that will hold. Be sure to use screws or rivets made of the same material as the gutter. Otherwise, the fastener or the gutter will likely corrode.

Wider downspouts are less likely to clog, and they handle intense rainfall better than smaller spouts. Going from 2-in. by 3-in. downspouts to 3-in. by 4-in. spouts doubles the flow potential, for very few dollars. Because elbows slow water, minimize turns and maximize vertical runs of downspout. If possible, install at least a short vertical section above any elbows so that the runoff enters the turns at a good clip. Fast-moving water can often clear small clogs.

Double the Usual Number of Hangers Where Ice Forms

In northern climates, ice often clogs gutters and downspouts. Iced gutters can contribute to ice dams or simply peel off the house from the weight of the ice. A few years ago, snow slides left my gutters a twisted pile of aluminum in the side yard. Here are a few techniques that can ratchet the odds in your favor.

Particularly on the north side of the house, slope the gutters as much as possible to give snowmelt its best chance to drain. Sloped gutters drain faster than level gutters, but long runs may look bad. Adding even a 1-in. drop on a long gutter will help. Install downspouts where they get at least a little afternoon sun that might melt accumulated ice. On the north side of a house, that may mean extending the gutter just beyond the end of the wall so that the downspout can drop down the western side of the corner.

Half-round gutters suffer less ice damage than those with rectangular sections because their shape tends to guide ice upward as it expands. Smooth-side downspouts that clog with ice are more likely to burst than those made of corrugated material. Corrugation has some ability to stretch. To combat the weight of accumulated ice and snow, the

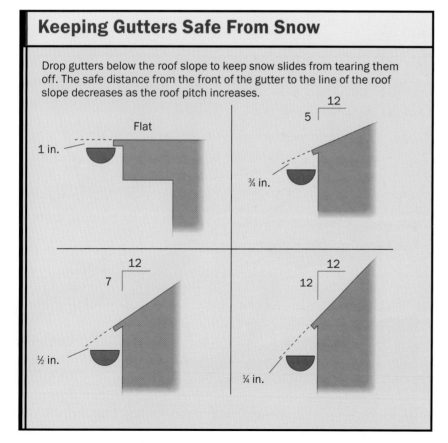

Keeping Gutters Safe From Snow

Drop gutters below the roof slope to keep snow slides from tearing them off. The safe distance from the front of the gutter to the line of the roof slope decreases as the roof pitch increases.

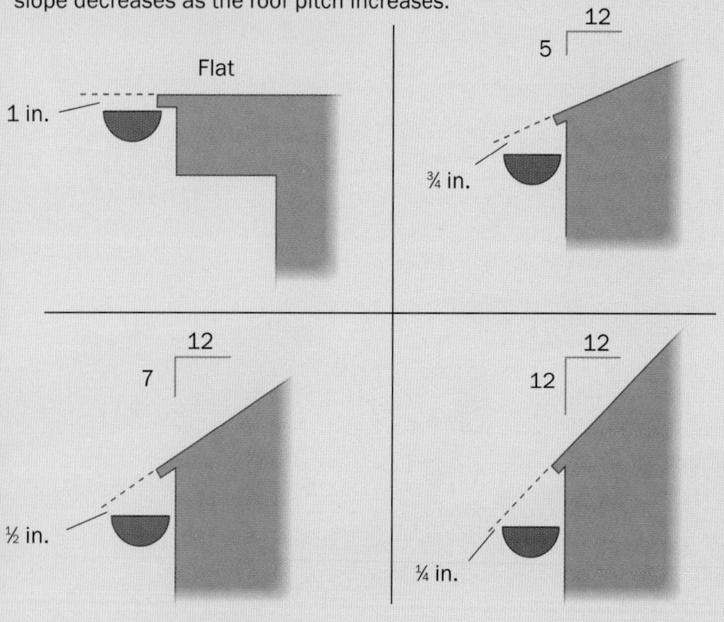

gutter must be supported properly. Where winters can be intense, space spikes or brackets every 1 ft. 6 in., or twice as frequently as normal. Another snow-country precaution: Install gutters so that the front lip is below the line of the roof slope (see the drawing on the facing page).

Runoff From One Roof to Another Can Damage Shingles

Upper roofs that drain onto lower roofs, such as on a porch that covers the front of a two-story house, may tempt you to decide against putting a gutter on the higher roof. Runoff from the upper roof will fall only to the lower roof, whose gutter will dispose of the rain. However, you'd be courting trouble. Even assuming you sized the lower gutter for the added runoff, concentrated water pouring onto the lower roof will erode the shingles and reduce their life span. Water that splashes onto the house's siding also can cause paint to peel and wood to rot.

The best solution is to put a gutter on the higher roof and to run its downspout to the ground. That solution may not be possible, for example, with a wraparound porch. In that case, if you don't mind the look, run a downspout over the lower roof and into its gutter. To drain the increased runoff, add an additional downspout to the lower roof's gutter.

If that's not feasible, you can simply drain the upper gutter's downspouts onto the lower roof. If you do that, though, be sure to drain it onto a splash block to avoid eroding the shingles. Even a cheap plastic splash block spreads out the flow. You can also custom-bend a splash block from sheet metal. Be sure to corrugate the bottom to disperse the water.

Roof valleys concentrate runoff, often to the point where it overshoots or splashes out of the gutter. The fix is to install a baffle, a sheet-metal corner that rivets inside the gutter's outside edge. Baffles typically rise about 2 in. higher than the gutter and block torrents from overshooting the gutter. Adding a downspout nearby will help water to drain quickly.

Hidden Hangers Install Quickly

The best way of affixing gutters to the house is subject to great debate. I haven't found a universal answer, but here's some information to help you choose.

To make gutters easier to clean, my ideal hanger wouldn't cross the top of the gutter (see the photo on p. 22). Brackets of this type cradle many half-round gutters and are available for copper K-style gutters. I couldn't find any for aluminum K-style gutters.

Most installers I spoke with use hidden hangers for aluminum K-style gutters. Some of these hangers are fastened with an integral galvanized-steel screw (see photo 2 on p. 30). This type clips to the back of the gutter and slips into the front channel. They're an installer's dream because they clip into place on the ground. There is no fumbling for fasteners on the ladder; the installer simply drives the preplaced screws home. Hidden hangers don't support the gutter bottom, though. Their strength depends on gutter rigidity. To be certain hidden hangers will hold up, it's best to use them only on heavier 0.032-in. gutter stock. And be prepared for the steel screws to corrode eventually.

Other brackets are nailed or, preferably, screwed to the fascia and support the gutter from below. A strap snaps across the top of the bracket, locking the gutter in place (see photo 3 on p. 30). These straps may be stronger than the hidden hangers because they support the gutter top and bottom. Some tie to the house with straps (see photo 1 on p. 30) that extend onto the roof, useful if there is no fascia, particularly on replacement jobs where rafter tails no longer offer sound fastening.

Most Common Gutter Hangers

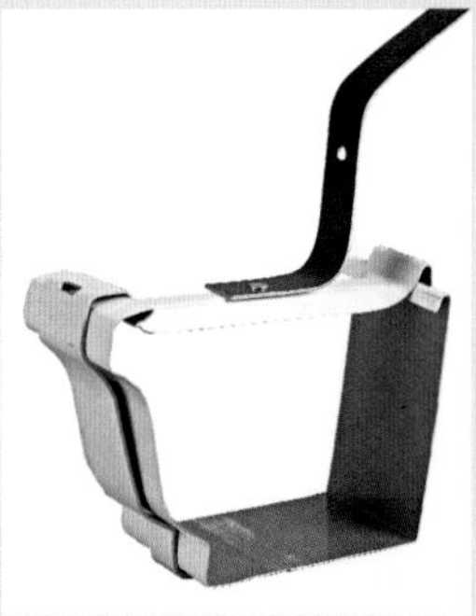

1. Strap nails to the roof. Supplied in a gawky one-piece configuration that is site-bent around the gutter, these hangers are handy when no fascia exists.

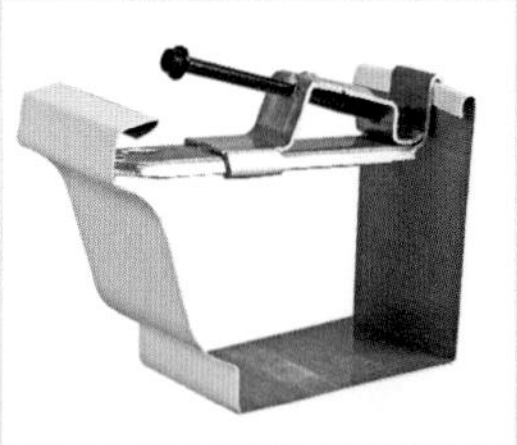

2. Hidden hangers are an installer's favorite. The finished gutter looks sleek, and because the hangers clip to the gutter on the ground, installation is a breeze.

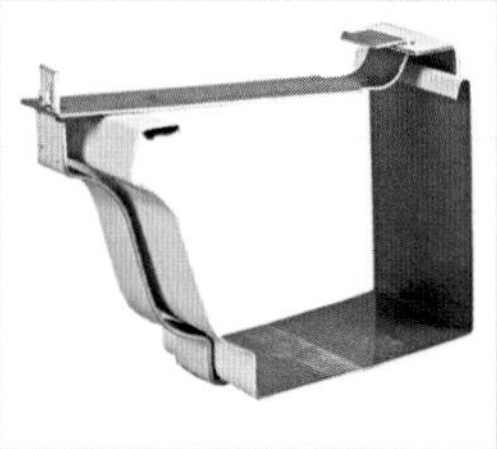

3. Snap-lock hangers help to keep the gutter from twisting. Installation takes more trips up the ladder than with the other hangers because the brackets are nailed to the house before the gutter is placed.

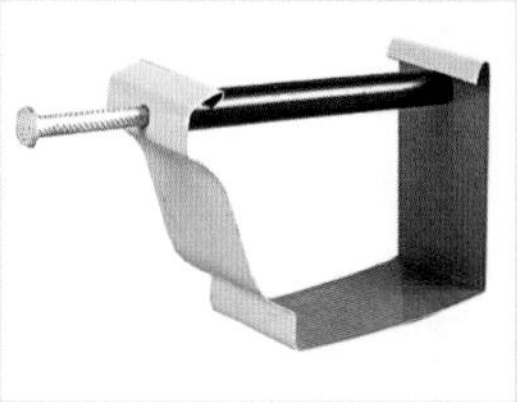

4. These devices are not your father's spikes and ferrules. Ribbed-aluminum spikes bite into the rafters or fascia and don't rust, unlike the smooth galvanized spikes common to early aluminum gutters. Ferrules keep spikes from crushing the gutter.

Brackets that support the gutter from below can be time consuming to install. Depending on how the brackets are spaced, you might be able to fasten three brackets for each trip up the ladder. Then, up the ladder with the gutter, and one more trip for every three brackets to snap on the top strap.

Spikes and Ferrules Are the Old Standby

Spikes and ferrules are probably the most controversial fastening method (see photo 4 at left) because older versions didn't work well and because they're tough to install. Spikes have evolved from the smooth galvanized versions that rusted and fell out of the fascia on my parents' house. The new spikes are ridged aluminum, hold pretty well and don't rust. They hold so well, in fact, that they're hard to pull without damaging the gutter. This holding power can be a problem should the gutter need to be removed, say, for painting the fascia.

Starting gutter spikes takes some skill. With the fingers of one hand splayed around the gutter, the installer holds the spike outside the gutter and the ferrule inside. Holding a hammer in his other hand, the installer delivers a sharp blow that sends the spike through the gutter's face into the ferrule. Aluminum spikes bend easily, and a misaimed hammer blow can ruin a gutter. A misfire like this would happen to me only on the last spike in a 50-ft. gutter. Unless the house has 2x fascias, common on some new homes but not on old, the spikes have to enter the rafters.

**Prices noted are from 1999.*

***Andy Engel**, a long-time managing editor at* Fine Homebuilding *magazine, is currently a senior editor at* Fine Woodworking *magazine.*

Draining Gutter Runoff

■ BY BYRON PAPA

To keep roof runoff away from foundations, my local code requires that water from downspouts drain at least 5 ft. away from the house. Many builders comply with the letter of this law by leading downspouts into buried 4-in. pipes that surface about 5 ft. from the house. Sometimes, though, 5 ft. away is uphill. I suspect that the water ends up back at the foundation.

On my houses, I drain roof runoff well away using inexpensive corrugated-plastic pipe. I tie the gutter drains to the footing drains beyond the house (see the top photo at right). I don't drain roof runoff directly into the footing drains because the added volume could overload them and flood the foundation. All the drains slope to daylight and terminate at one or two points (see the bottom photo at right).

Foundation and gutter drains join beyond the house. Otherwise, water from the gutters might flow out of the perforated foundation drains, flooding the footings.

Slotted caps bar vermin from drains. Runoff from the entire roof concentrates here, so cobbles are important to disperse the flow and to reduce the chance of washout.

Planning Simplifies Installation

Before beginning, I have my gutter installer visit the site to mark the downspout locations. Next, I sketch the drain layout. This sketch relieves me of having to think much in the flurry of running the pipes and helps in creating a materials list. I note any

TIP

The rule of thumb is that each square inch of a downspout's section can drain 100 sq. ft. of roof, and that factor applies to drain pipes, too.

changes that have been made to the original plan during installation and revise a copy to give to the homeowner.

I use unslotted 4-in. pipe for the gutter drains to be sure that no water from them ever wets the footings. This pipe comes in rolls of 50 ft., 100 ft., and 250 ft.; these lengths minimize waste and allow me to limit underground joints to those needed for tee- or wye-fittings. The pipe is flexible enough that elbows aren't needed. Joints are openings for roots to enter and clog pipes, so the fewer joints, the better.

Smooth-wall PVC pipe is an option that might allow better flow than corrugated pipe. I don't use it because it's more expensive and because it comes in 10-ft. lengths that must be spliced underground.

Downspout Connections Double as Clean-Outs

I begin the gutter drains with vertical sections at the two downspouts where the final grade will be highest. From these points, I bend the pipes horizontal and run one around the house clockwise, the other counterclockwise. I let the ends stick above grade and cut them off when the gutter installer runs the downspouts. Commonly available adapters fit the downspouts to the pipe (see the photo below). The downspouts detach easily, providing access to clean the drains of clogs.

Additional downspouts can tap into these pipes with wyes, but I'm careful not to overload a single pipe. One 4-in. pipe will handle runoff from one 3-in. by 4-in. downspout or two 2-in. by 3-in. downspouts. The rule of thumb is that each square inch of a downspout's section can drain 100 sq. ft. of roof, and that factor applies to drain pipes, too. Following that rule, one 4-in. pipe can drain about 1,200 sq. ft. of roof. I try to pitch the drains ¼ in. per ft., but on flat lots, there isn't always enough elevation difference to make this degree of pitch possible.

Transition neatly joins downspout to drain. Although the downspout could feed directly into the drain, transitions cost little and keep out debris that could lead to clogs.

Gutter and Footing Drains Connect Away From the House

I build most of my houses on crawlspace or slab foundations, so the foundation and gutter drains often run very close together (sometimes even side by side). I gather all the pipes together in a common trench leading to their outfall, where they empty to the surface. In this trench, I tie the footing drains into the gutter drains. This plan reduces the number of pipes at the termination (see the drawing on the facing page). A caution: If the house has underground utilities, I check that their installation hasn't damaged these drains.

I've considered stepping up to 6-in. pipe instead of running separate 4-in. pipes. I decided against this option because 6-in. pipe is stiff and difficult to handle. And although one 6-in. pipe will handle slightly more flow than will two 4-in. pipes, it's also much more expensive than two 4-in. pipes.

An average three-bedroom house usually ends up with four pipes at the outfall point. I use two outfalls to avoid concentrating the flow from bigger houses. The pipes terminate in a bed of pebbles that's downhill and

Subsurface Drains Take Roof Runoff Away From the House

Codes call for drains to carry the water at least 5 ft. from the house, but farther is better. Gutter drains can tie to foundation drains far enough away from the house that roof runoff won't back up and flood the foundation.

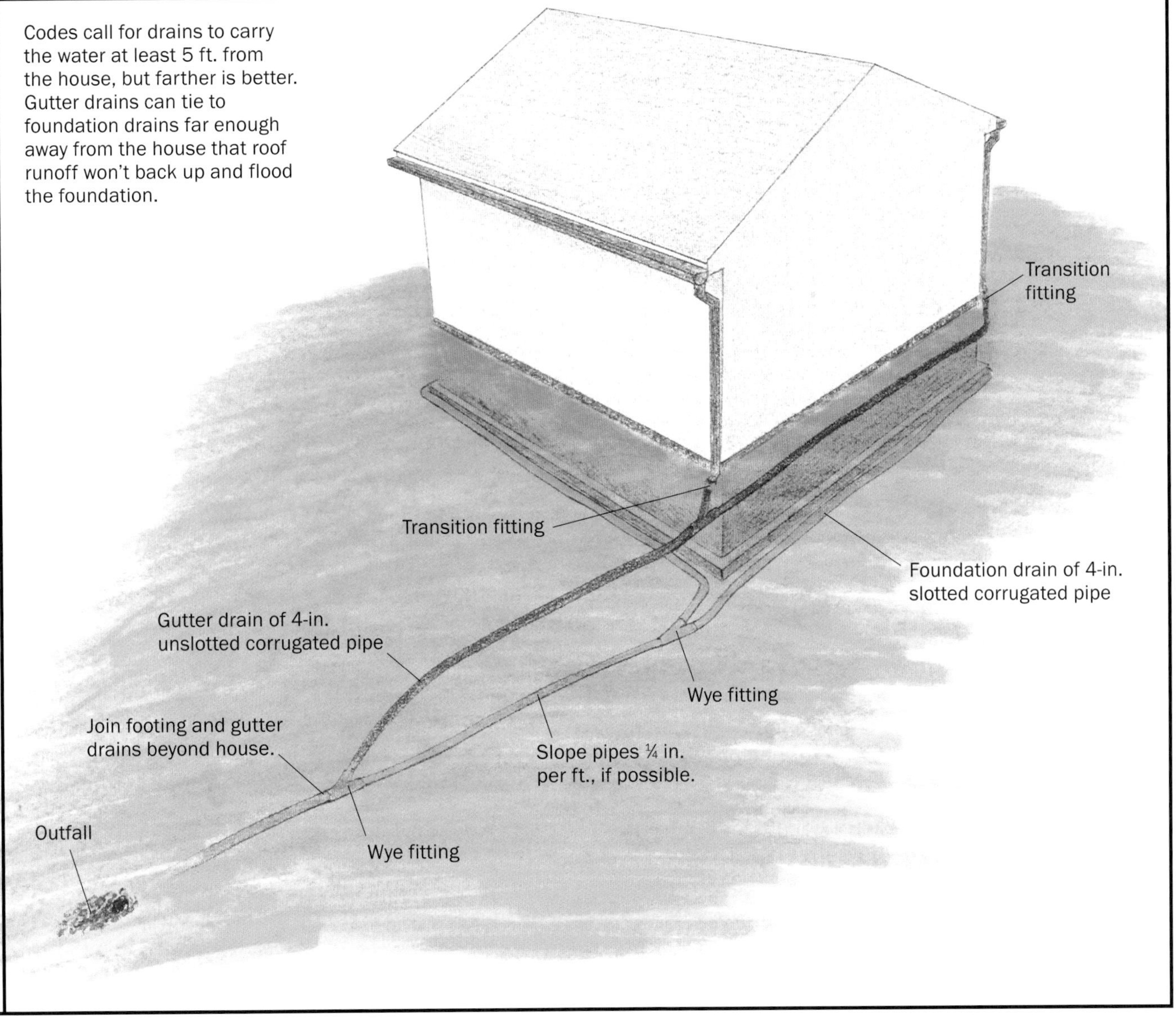

that is at least 10 ft. from the house (see the bottom photo on p. 31). I'm careful not to place the outfall in a location where it can flood a neighboring lot.

Sometimes, when site conditions preclude an adequate drainage swale around a house, I'll place a catch basin in the poorly drained area and tie it to the gutter drains. My local home center sells plastic catch basins that accept the pipe I use.

When all the pipes are laid, I top them off with 2 in. or 3 in. of gravel, then fold filter fabric over the gravel toward the foundation wall. The remaining backfill pins down the fabric. The foundation and gutter drains for a typical 2,500-sq. ft. house cost about $800* for pipe, gravel, fabric and labor. Most houses get foundation drains anyway, and the gutter drains account for less than half this cost.

**Prices noted are from 1999.*

Byron Papa *is a building and remodeling contractor in the Durham/Chapel Hill, North Carolina, area who works with green building methods and materials.*

Roof Flashing

■ BY SCOTT McBRIDE

I once heard about a man with a perpetually leaking roof. He said that he couldn't very well fix it in the rain, and that when the sun was shining, he didn't need to. Most likely, the problem could have been avoided if the roof flashing had been done right in the first place.

Although the underlying principles are the same for both sidewall and roof flashing systems, roof flashings are more severely tested; rain strikes a roof directly, and the sun bakes it throughout the day. Good sidewall flashings can extend the life of a structure, but roof flashing keeps the living room from turning into a rice paddy.

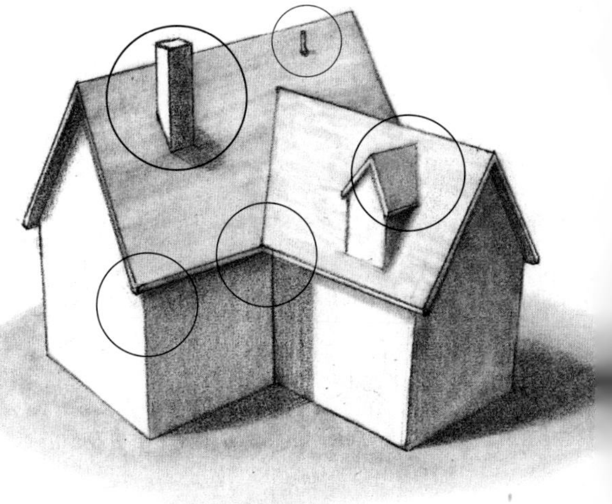

EXTRA PROTECTION AT EDGES, SEAMS, AND PENETRATIONS.
Flashing is like long underwear: an extra layer of protection against the weather. The illustration above highlights critical flashing areas on a typical roof, and in the drawings that accompany this article, the illustration serves as an icon to identify where on the roof a particular flashing is located.

Edge Flashings Go On First

The first flashing to be installed on a typical roof is the edge flashing (see the drawings on the facing page). Metal roof edge, also known as drip edge, protects the edge of the roof sheathing from water splashing out of the gutter and from the moisture present in gutter debris. It also fends off moisture that gets drawn under shingles by capillary action. Finally, it makes for a neat appearance, especially along the rake.

In northern climates, where ice dams are a problem, the use of self-adhesive bituminous sheeting at the eaves is becoming almost routine. Typically, one 3-ft.-wide sheet of bituminous membrane is installed at the eave with the rest of the roof covered in overlapping courses of roofing felt. Nail the drip edge to the deck along the eaves and nail it over the roofing felt and bituminous sheeting along the rakes.

Metal roof edge generally is available in one of two profiles, one that resembles an L

Eave Flashing

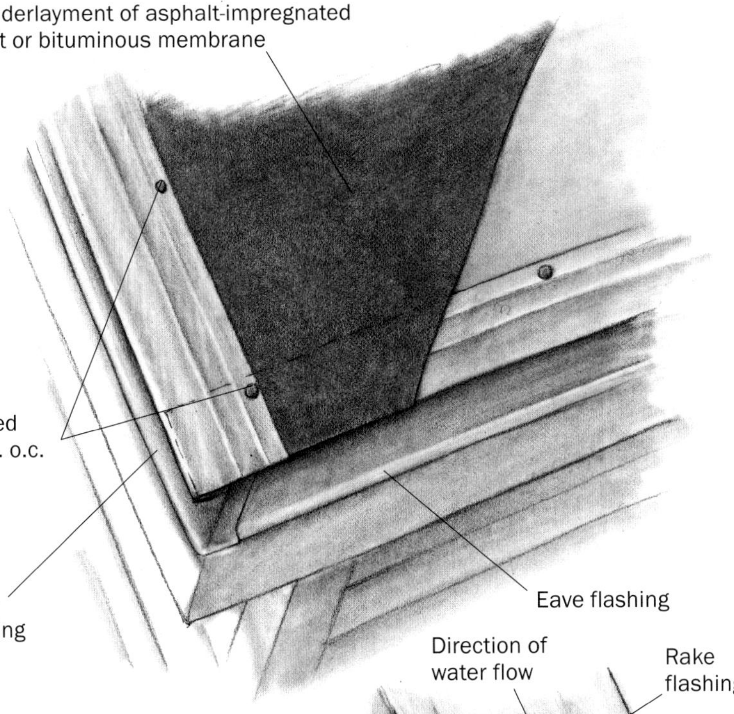

ROOF-EDGE FLASHING BRIDGES THE GAP BETWEEN ROOF AND FASCIA
Roof-edge, or drip-edge, flashing is installed along the eave first, directly over the sheathing. Then roofing felt goes on. Finally, drip edge is installed along the rake. The nails are placed near the upper lip of the roof-edge flashing to ensure the nail holes will be covered by other roofing materials.

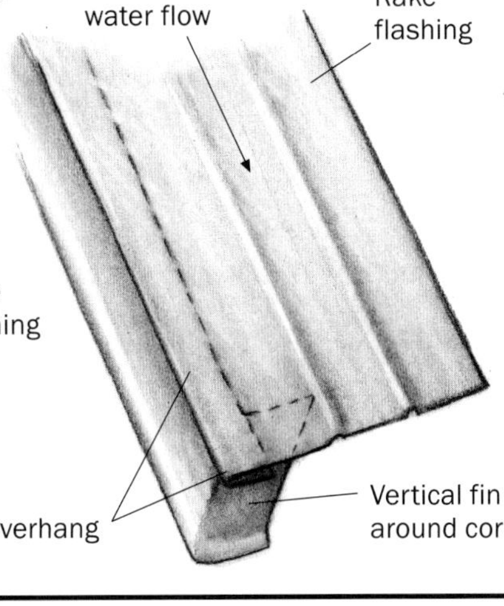

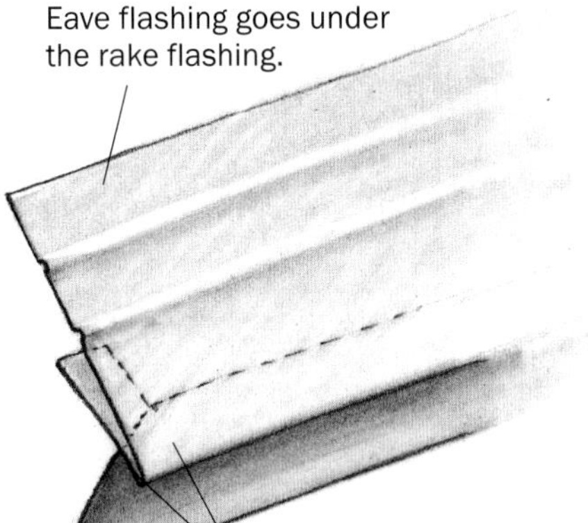

INTERLOCKING ROOF EDGE KEEPS THE WATER OUT.
To join sections of T-shaped drip edge at the corner of the rake and eave, the author cuts and bends the two pieces as shown. The eave flashing is installed first, and then the rake flashing is slipped over it. Water running down the rake is directed over the eave section and off the roof.

and another that looks like a lopsided T. I like the T-profile because it supports the shingle overhang. Drip edges are sometimes available in a variety of widths; if you've got a choice of materials, wider is better than narrower. With the wider stock, nails can be placed slightly higher on the roof, thereby reducing the risk of water damage to the sheathing.

I nail roof edge directly to roof sheathing, about 12 in. o.c., placing the nails as high as possible without weakening the flashing. At the corner where the rake meets the eave, I run the eave flashing first (see the drawings above). The small overhang that sticks out past the eave is allowed to project a bit past the rake as well. When the rake flashing comes down, it captures the overhang. A slit along the crease of the rake flashing then allows its vertical fin to bend around the corner.

There are two types of vented eave flashing available (see the drawings on p. 36). They can be used where traditional soffit

Vented Roof Edge Performs Two Functions

Vented roof edge serves as a drip edge and a roof vent. Air passes through vent slots in the material and then into a gap between the fascia and the roof sheathing. The material eliminates the need for soffit vents and can be installed on a house with little or no soffits.

COMBOVENT

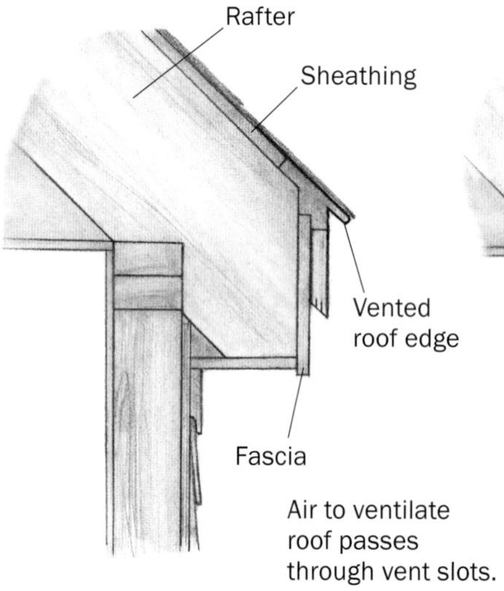

STANDARD DRIP-EDGE VENT

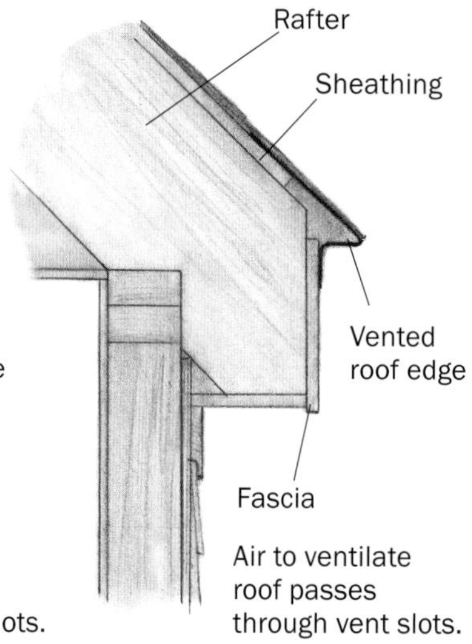

vents won't work. One of these, the ComboVent (also sold as the SmartAir® intake vent) is a new product made of PVC and is larger than typical metal drip edge. The bottom leg, about 4 in. long, is held away from the fascia, allowing air to pass behind the gutter and under the roof sheathing.

Valley Flashing: Open or Closed?

Good valley flashings are critical because roofs drain into the valleys. Most roof leaks can be traced to faulty valleys. Valley flashings fall into two categories: open and closed. Open valleys have exposed flashing, and closed valleys do not. The chief problem with closed valleys is that they tend to collect debris. Without maintenance, the debris can cause water to back up under the shingles. Debris also holds moisture, which leads to the premature decay of both roofing and flashing.

Synthetic Flashings Are Cheap and Easy to Use

Metal still figures prominently as a roof-flashing material, but in recent years some inexpensive and highly workable synthetic alternatives have come on the market.

The first synthetic-flashing material was asphalt-impregnated felt, which is known universally as tar paper. It's still used in low-stress situations where the entire flashing is protected by another material, such as when it's used around windows.

Felt breaks down quickly when it's exposed to the weather. Heavier felts, known as roll roofing or 90-lb. roofing, can be used for exposed-valley flashing. A gravel coating on roll roofing gives it some protection from the damaging effects of sunlight. Valleys flashed with roll roofing can be color-coordinated with a roof's asphalt shingles. Unfortunately, roll roofing has a short life span compared with metal and is highly susceptible to punctures. As with roll roofing, synthetic materials that were recently developed as roof membranes have been carried over into flashing work. EPDM is a synthetic rubber sheet that can be solvent-welded. Modified bitumen is a sheet material welded with a flame. The ability to join separate pieces of these materials makes them a rival of copper, especially in low-pitch situations where overlapping alone will not create an adequate water barrier.

For example, built-in gutters can now be relined with these materials at a fraction of the cost of relining them with copper. Prefabricated rubber fittings are an added time-saver. For example, you can buy a flashing boot for 4x4 posts that solvent-welds to an EPDM deck membrane.

Closed Valleys Are Easy With Asphalt Shingles

Closed valleys leave no metal or other flashing material exposed, but they are more likely to collect debris than open valleys. Two ways of constructing closed valleys are shown below. Of the two, woven valleys take longer to construct because you have to work both sides of the roof at the same time.

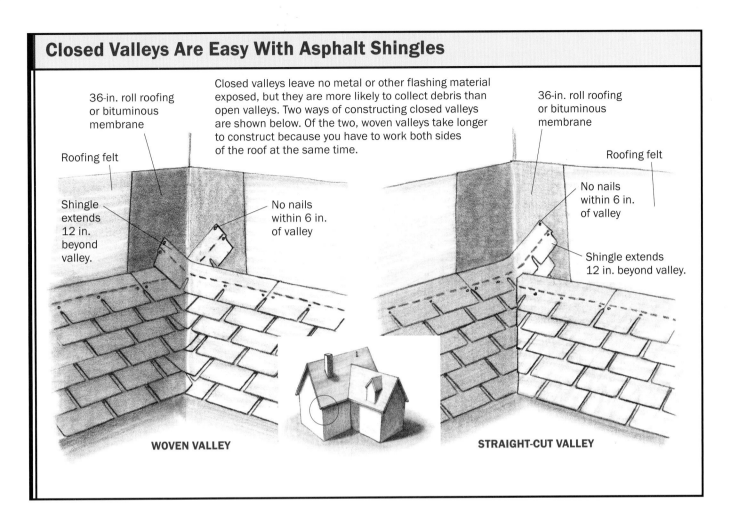

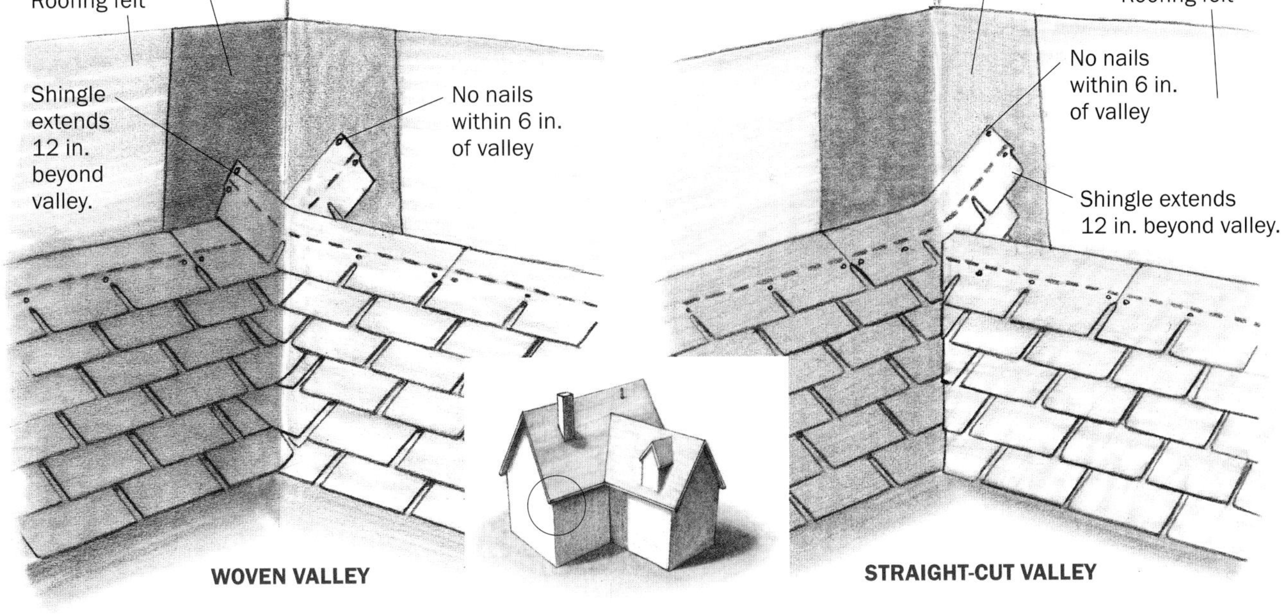

When you're working with an asphalt-shingle roof, closed valleys are easy to achieve. The shingles are flexible, so they can act as their own flashing. The two types of self-flashing closed valleys are woven and straight cut (see the drawings above). When roofing with an inflexible shingle, such as wood, slate, or tile, closed valleys require metal flashing. The flashing is in the form of concealed metal "shingles" woven into the roofing one course at a time, like step flashing.

Open Valleys Are Functional

Open valleys, commonly made of galvanized steel, aluminum, or copper, are quick to install, and they drain well. I install valley flashings over the roof underlayment (felt or bituminous membrane). The underlayment cushions the flashing against any protruding nail heads and flashing. Each course of underlayment is extended across the valley and onto the adjoining roof. Successive courses are woven together in this fashion, creating a double thickness of underlayment in the valley that serves as a backup in case the flashing fails.

On shallowly pitched roofs, metal flashing will lay nicely in an open valley without a crease. With steeper pitches, I crease the middle of the valley on my brake, a tool I use for bending flashings. When there's more water coming down into a valley from one side than from the other, a V-crimp down the middle is recommended. This ridge keeps the heavier runoff of one side from flowing over the middle of the valley and up under the shingles on the opposite side. To fasten an open-valley flashing, I drive roofing nails just beyond the edge of

TIP

To avoid galvanic corrosion, it's wise to use the same materials for flashings that will be in contact, such as valley flashing and roof edge.

the flashing at 24 in. o.c. The head of the nail holds down the sheet metal but allows it to expand lengthwise. I through-nail only at the top end to keep the flashing from slipping downhill.

To avoid galvanic corrosion, it's wise to use the same materials for flashings that will be in contact, such as valley flashing and roof edge.

Cleats Allow Valley Metal to Expand

To ensure that valley flashing doesn't buckle as it expands in heat, you can also attach it to the roof with cleats instead of nails. Cleats hook over a fold turned along the edges of the valley (see the drawing below). The fold is free to slide in the cleat, thereby accommodating expansion. The fold also turns back water that reaches under the shingles, and it lets a little air circulate between roofing and flashing.

Limiting the length of each valley sheet to 10 ft. also will help to control expansion. If more than one sheet is required in a valley, the upper sheet should overlap the lower sheet by about 12 in.

At the bottom of an open valley, where the valley meets the eave, the valley should be trimmed so that it overhangs the roof slightly (½ in. to 1 in.). This way, water will be carried past the inside corner of the roof. Take care not to extend the valley too far, though, or the runoff could flow past the gutter.

Typically, the valley flashing is run up to and cut off in line with the ridge. The end of the valley flashing then is covered by the ridge-cap shingles or the ridge vent.

Metal Valley Flashing Will Expand and Contract

Allow for this movement by limiting valley pieces to about 10 ft. (overlapping them 12 in.) and by attaching metal valleys either with cleats or with nails along the edge of the valley, trapping it beneath their heads.

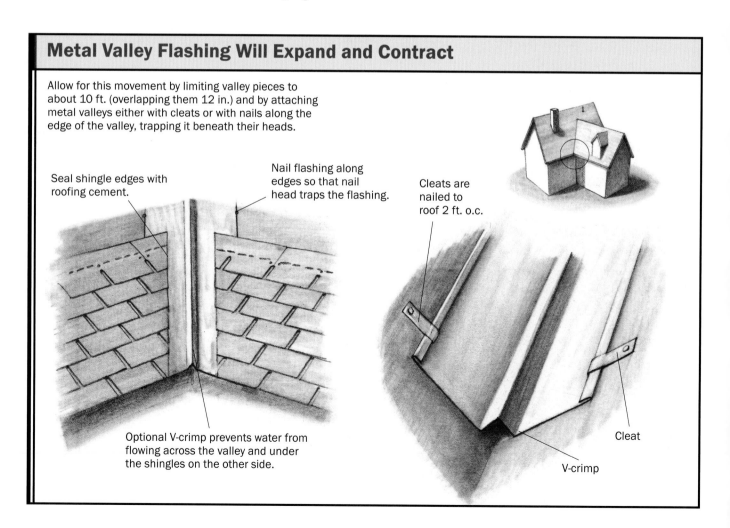

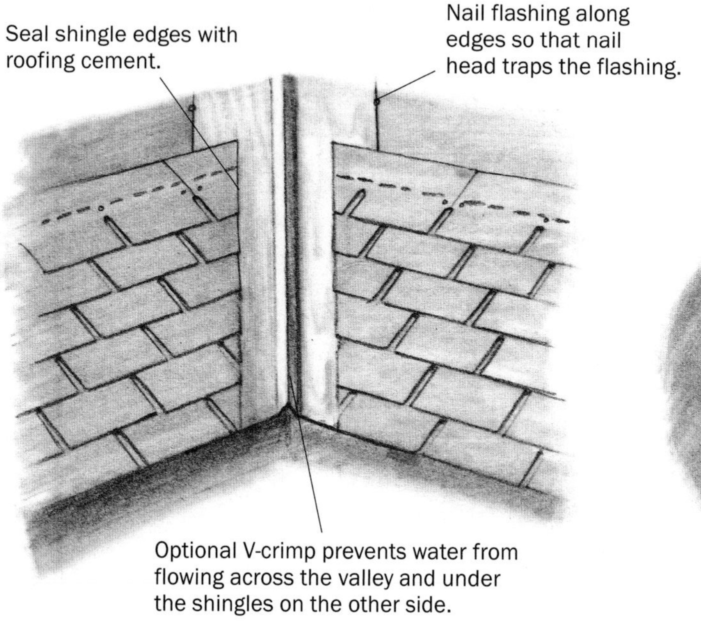
Seal shingle edges with roofing cement.
Nail flashing along edges so that nail head traps the flashing.
Optional V-crimp prevents water from flowing across the valley and under the shingles on the other side.

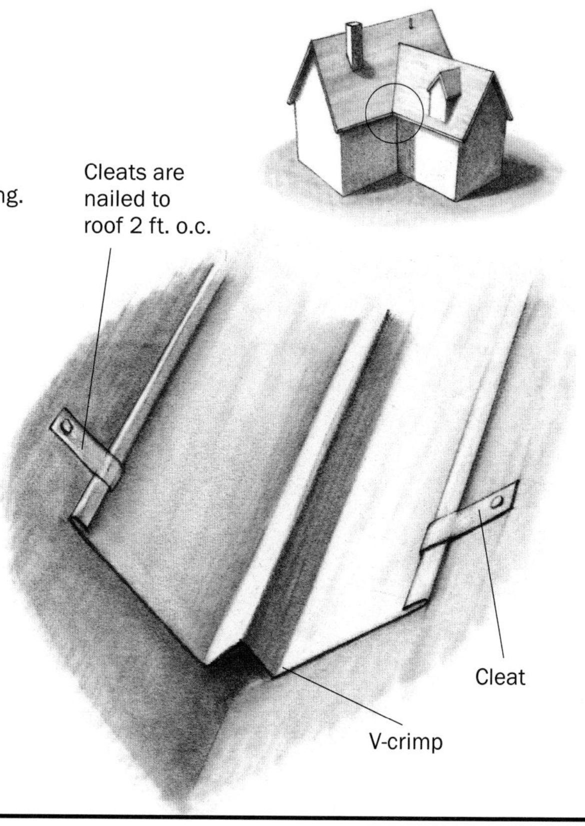
Cleats are nailed to roof 2 ft. o.c.
Cleat
V-crimp

The same principle applies at the intersection of two valleys, such as at the back of a dormer. One way to handle this condition is to run both pieces of valley flashing long, past the intersection of the dormer ridge and the main roof (see the drawings at right). The flashing from both valleys is cut down the middle, at the crease. The four sections of flashing are then woven together, with two halves lying flat and overlapping on the main roof. The other halves meet on the ridge of the dormer and are soldered or folded and crimped together.

Mineral-surfaced roll roofing, also known as 90-lb. roofing, can be substituted for metal in open valleys. To increase strength, a double layer is used. The first layer is at least 12 in. wide, laid gravel side down. The second layer is at least 24 in. wide, laid gravel side up. The two layers are laminated together with a troweled-on roof cement. Despite this precaution, boot heels will easily puncture a roll-roofing valley.

After installing an open valley, I snap chalklines indicating the lines at which the overlying shingles will be cut. The exposed width of the valley starts at about 6 in. at the ridge (3 in. per side) and gets wider as the valley descends to accommodate the increasing volume of runoff. The recommended taper is ⅛ in. per ft. For instance, a valley 16 ft. long would measure 6 in. wide at the ridge and 8 in. wide at the eaves.

When shingling at a valley, keep the nails at least 2 in. back from the line where the shingles are trimmed. If this leaves a shingle sticking out more than 6 in. from a nail, secure it with a small dab of roof cement.

Intersecting Valleys Are Cut and Woven Together

When flashing from two valleys intersects, such as at the back of a dormer, the flashing can be cut down the middle, along the crease, and the four sections or tabs can be woven together.

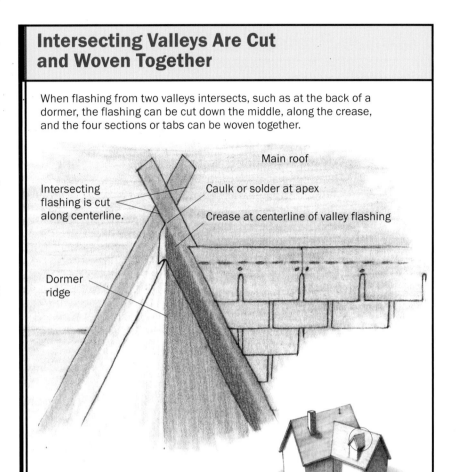

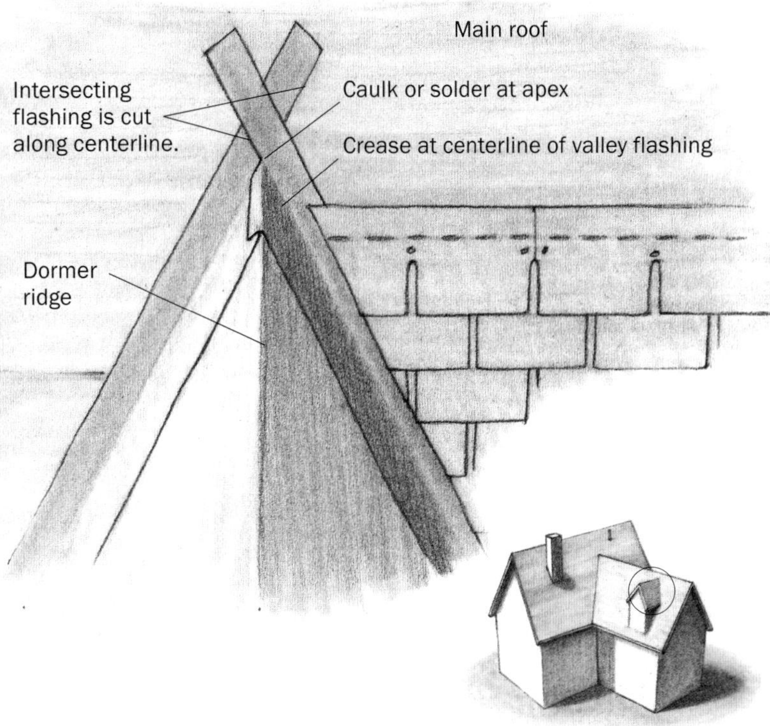

Chimney Flashing Is Folded Into Masonry

Chimneys are flashed in much the same way as dormer walls. The difference lies mainly in what covers the upturned sides of the apron, or base, flashing and the step flashings. Whereas dormers have siding that comes down over the flashing, masonry chimneys use counterflashing, which I'll discuss later.

The first piece of chimney flashing to go on is the apron flashing, which laps at least 4 in. over the shingles below and at least 12 in. up the vertical face of the chimney. Then step flashings are installed along the sides of the chimney, woven in with the adjoining shingles.

What happens next depends on the chimney's location on the roof. When a chimney is situated at the ridge, the step flashings simply culminate with a top step that straddles both sides of the roof. But when a chimney is built downslope from the ridge, a cricket is employed (see the drawings on p. 40). A cricket, or saddle, is a

Chimney Flashing

CHIMNEY FLASHING BEGINS AT LOWEST POINT
Typically, the first piece of flashing to be installed on a chimney is the lowest piece, the apron, or base, flashing. Folded step flashing then is installed along the side of the chimney, one piece for each course of shingles.

COUNTERFLASHING IS WOVEN INTO MASONRY
The lower flashings on all four sides of the chimney are covered with counterflashing. The upper edge of the counterflashing is bent at a right angle and inserted about an inch into the mortar joints between the brick or stone of the chimney.

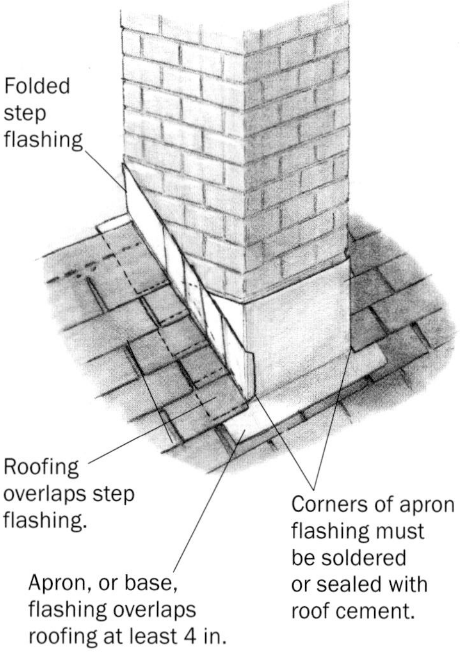
Folded step flashing
Roofing overlaps step flashing.
Apron, or base, flashing overlaps roofing at least 4 in.
Corners of apron flashing must be soldered or sealed with roof cement.

CRICKET BRIDGES THE VALLEY BETWEEN CHIMNEY AND ROOF
If the chimney comes through the roof below the ridgeline, a soldered sheet-metal cricket can be installed on the uphill side of the chimney. The valley flange on the cricket is covered by the roofing. At the corner of the chimney, the flange wraps around and over the folded step flashing.

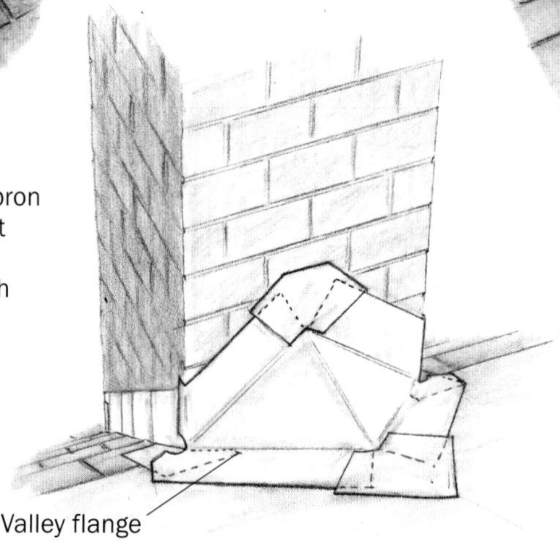
Valley flange

Counterflashing

miniature gable roof on the upslope side of the chimney that diverts water around the chimney. Some crickets are big enough actually to be shingled, but most are covered with a metal skin bent and soldered to fit the slope of the roof. A metal cricket is part valley flashing, part step flashing, and part metal roof.

The last step, whether a chimney is at the ridge or downslope, is the installation of counterflashing (see the drawings above). Counterflashing overlaps all the lower flashing pieces, including the apron or aprons, the step flashings along the side, and the cricket, if there is one. The top edge of the counterflashing is bent at a right angle and let into successive mortar joints in the chimney.

For a stone chimney, the irregularity of the material makes it tough to let in the step flashing neatly. To provide straight joints for step flashing, a mason can substitute brick for stone at the roofline. Brick is hidden in the attic space and concealed from the outside by step flashing.

A good mason leaves the appropriate mortar joints unpointed, making it easy for the roofer to install counterflashing. To pin the folded-over lip of the counterflashing in the mortar joint, I use rolled-up strips of lead flashing. The plugs are mashed into the joint to hold the flashing in place, and the joint is either caulked or pointed with mortar.

In remodeling work it is sometimes necessary to flash a new roof where it dies into an existing chimney. Rather than chisel out mortar joints to receive new flashing, I prefer to cut a continuous kerf parallel to the new roof. I use an abrasive blade mounted in a circular saw for this operation. The top edge of a continuous counterflashing can be folded into this kerf, which is then sealed with caulk.

Through-Pan Chimney Flashing Stops Moisture

Although most of the water striking a chimney runs off its sides, some moisture is absorbed into the porous masonry. This is especially true of stonework, with its wide, irregular mortar joints. This moisture can migrate far, eventually finding its way into the house. The solution is through-pan flashing, an expensive but effective alternative to regular chimney counterflashing.

First, the mason brings the chimney just above the roofline. Base flashing and step flashing are installed. A sheet of copper or lead flashing with a hole in the middle for the flue then covers the whole chimney. The masonry then continues upward, built on top of the flashing. Through-pan flashing creates a complete break in mortar bond just above the roofline. There is debate among masons about the wisdom of doing this. Some argue that mortar has little tensile strength anyway, and that the mass of a through-flashed chimney will hold it in place. Others feel that the mortar bond is an indispensable part of a chimney's vertical

Flashing Skylights in a Metal Roof

I recently installed skylights on my shop building, which has prefab metal roofing, sometimes referred to as "ag panels" (agricultural panels). I wanted to shed the runoff from the skylight safely beyond the panel ribs flanking the opening. To achieve this, I first built an auxiliary curb of 2x4s around the opening, which raised the level of the skylight jamb above the panel ribs. I flashed and counterflashed the curb with a custom-made galvanized flashing that extended beyond the ribs (see the top photo below), and then mounted the skylight with its own flashing on top of the curb (see the bottom photo).

Vent-Pipe Flashing Goes On Quickly

Once the area around the pipe has been shingled, the metal flange and rubber or neoprene gasket is slipped over the pipe and bedded in roof cement. The next course of shingles is notched to fit around the flashing, leaving 1 in. of clearance.

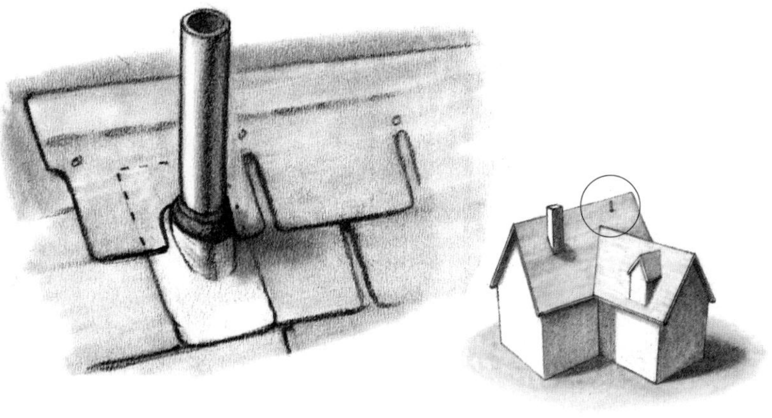

integrity. Common sense suggests that through-pan flashing might not be a good idea for tall, unsupported chimneys. On the other hand, chimneys that extend only a few feet above the ridge aren't going to blow down, through-pan flashing or otherwise.

Follow the Directions When Flashing Skylights

Skylights are furnished with their own flashing kits, so the manufacturers' directions are your best guide to installation. For sloped roofs, the better skylights employ step flashing rather than continuous side flashing. The trouble with continuous side flashing is that water spreads out laterally as it flows downhill. As the length of a skylight from top to bottom increases, so does the likelihood that water will find its way out past the edge of the flashing and into the structure. Step flashings prevent this by shunting water out onto the roof surface. This system ensures that water will be promptly shunted onto the roof surface, as with sidewall step flashing.

Sloped-skylight flashing kits typically consist of a few basic elements. An apron piece surrounds the bottom of the unit, extending out over the shingles below. Step flashings run up the sides, woven in with the shingles. A rubber or metal U-channel covers the top edges of the step flashings. Finally, a wraparound shroud covers the upslope side of the unit with a flange that slips under the shingles above.

From Lead Boots to Rubber Sleeves

In bygone days, the most common flashing for plumbing vent stacks was a lead boot. The boot fit over the stack, and its top edge was beaten down over the rim of the pipe. The surrounding flange of the boot lapped over the roofing below and slipped under the roofing above. The lead boot has been supplanted largely by a metal flange, usually aluminum, attached to a neoprene or rubber gasket (see the drawing at left). The rubber stretches around the pipe to accommodate different roof pitches. Although not as durable as lead, the newer hybrid boot gives good service at minimal cost. When replacing an asphalt-shingle roof, it pays to inspect the stack flashings and replace them if the rubber has deteriorated.

To install a stack flashing, first bring the roofing up to or just past the stack. Slip the boot over the stack and bed it in roof cement. Then continue with the roofing so that the lower half of the flashing is exposed while the upper half is covered. I trim the overlapping shingles at least 1 in. away from the base of the boot so that pine needles and such can be washed away.

***Scott McBride** is a contributing editor of* Fine Homebuilding *magazine. His book* Build Like a Pro: Windows and Doors *is available from The Taunton Press. McBride has been a building contractor since 1974.*

Flashing a Chimney

■ BY JOHN CARROLL

Laying up perfect steps. A variety of clamps allows the author to set the flashing at the same time he builds the chimney. The welding clamps secure counterflashing yet leave room for upper courses of brick.

In the early 1970s, my father and a friend went into the roofing business. The friend supplied the capital, and my father, who had six sons, supplied the labor. Being the most particular of the six boys, I was given the task of flashing chimneys. My work was good enough to justify my father's confidence, but invariably, the finished product was a hodgepodge of incongruent shapes and oozing tar—nothing to be proud of.

In the intervening quarter-century, I've built as many chimneys as I've flashed. As I've grown from a schoolboy roofer to a seasoned builder and mason, experience has taught me that the best way, visually and structurally, to flash a chimney is to do so at

An upper piece of flashing always laps over a lower one. To ensure that every drop of water is deposited on the surface of the roof, the lower piece of flashing always laps over a shingle.

the same time the bricks are being laid. Being skilled in both trades is convenient, but it's not a requirement. If plumbers and carpenters can rough in a bath together, a mason and a roofer working in a spirit of cooperation can easily do this job.

Chimneys Need Double Protection

Chimney flashing looks complicated because it involves two distinct components: base flashing and counterflashing. The base flashing covers the joint between the chimney and the roof, ensuring that all the water that flows down the roof is channeled back onto the surface of the shingles. The counterflashing laps over the top of the base flashing to ensure that any water that runs down the chimney is channeled to the outside of the base flashing.

Chimneys break through the surface of roofs in every conceivable place: at the eaves, on the ridge, along the rake, and almost everywhere in between. As different as these configurations might be, the strategies for directing water to move harmlessly around a chimney are basically the same. Whenever I'm installing roofing or flashing, I always visualize the flow of water. An upper piece of flashing always laps over a lower one. To ensure that every drop of water is deposited on the surface of the roof, the lower piece of flashing always laps over a shingle.

Flashing Bends Are Made in Advance

I certainly didn't invent the idea of installing the flashing as the chimney is being built. Bricklayers in New England, where lead flashing is popular, often bed counterflashing in the mortar joints, then bend the flexible metal upward until the roofers come along to fit the base flashing underneath. There is nothing wrong with this method, but I prefer to install all the flashing at the same time. This technique gives me complete control over the flashing process and allows me to use better-looking material.

Besides lead, there are many kinds of sheet metal that roofers use to flash chimneys. I prefer copper. Standard 16-oz. copper bends crisply and has enough body to stay straight and smooth. Unlike aluminum, it doesn't corrode when embedded in wet portland cement, and unlike steel, it never rusts. To my eyes, copper looks great alongside brick or stone, and like those materials, its appearance improves with age. Handsome, durable, exuding quality, copper is truly the Irishman of flashing materials.

Flashing a chimney involves a lot of cuts and bends that must be made on site, but I prefer to have the rough shapes formed ahead of time in a sheet-metal shop. Getting these pieces fabricated in a shop means the bends are crisp and the material is straight and smooth. Besides just looking good, straight metal makes for a tight fit, and the tighter the fit, the less chance that water will seep into the house.

Bricks Are Laid to Accommodate the Flashing

Although the base flashing carries the most water, the counterflashing is most visible. I want my counterflashing to look as even as a good set of stairs and to hug the chimney securely. Before I bring the chimney up through the roof, I work out a layout for the counterflashing, then I set the bricks to conform to the layout. I make sure there is a vertical joint in the brickwork at every step in the counterflashing, which allows me to bed the vertical legs of the flashing in the mortar joints along with the horizontal legs.

On moderately pitched roofs (up to about 8-in-12), I generally step up the counterflashing one course of brick at a time. On steeper pitches (up to 16-in-12), I step it two courses at a time, and on very steep pitches

Uniform steps of brick make for perfect flashing. The author uses a 3-in.-wide board as a gauge to lay out a consistent set of steps. The next course of brick begins where the board meets the top of the brick.

Base flashing turns the corners of the chimney. The author scribes the edges of the chimney on the backside of the flashing (photo left). After cutting and bending both ends (photo below) to match the contours of the chimney (see the drawing on p. 46), he anchors the base flashing by driving one copper roofing nail through each outside edge.

(over 16-in-12), I'll make it three courses at a time. On this 1830 cottage, the roof pitch was an oddball (but not uncommon) 6⅝-in-12, so I knew that the counterflashing would step up with each course of brick. But I still had to make sure that those bricks stepped up consistently with the roof.

As I brought the chimney up through the roof, I laid two full courses of brick above the roofline and stopped. Then I used a 3-in.-wide board as a gauge to lay out a consistent set of steps in the counterflashing (see the photo above). I set the board on the roof and marked where the top of the board intersected with the top of the bricks. At that point, I began the next course of bricks. I repeated this process three times and ended up with four uniform steps marching up the roofline. Because this chimney emerged through the ridge, I had to adjust the bricks on both sides of the chimney.

Apron Flashing Is a Copper Shingle That's Folded in the Middle

Before I began the base flashing, I ran the shingles up the roof until the tops of the cutouts were within 5 in. of the bottom of the chimney. The front piece of base flashing (also known as the apron) would essentially serve as the next shingle (see the top photo above). To make up the apron, I'd had the shop bend a 10-in.-wide strip of copper down the middle so that the top leg could extend 5 in. up the chimney while the bottom leg extended 5 in. down the roof. The shop also had cut the apron 8 in. longer than the 24-in.-wide chimney. This cut allowed a 4-in. overlap on each end to wrap the corners of the chimney (see the bottom photo above). Using tin snips and a pair of hand seamers (Malco Products, Inc.; 800-596-3494), I shaped the apron as shown in the drawing on p. 46. I bent the ends a little more than a true right angle to ensure that

Getting Off to a Good Start

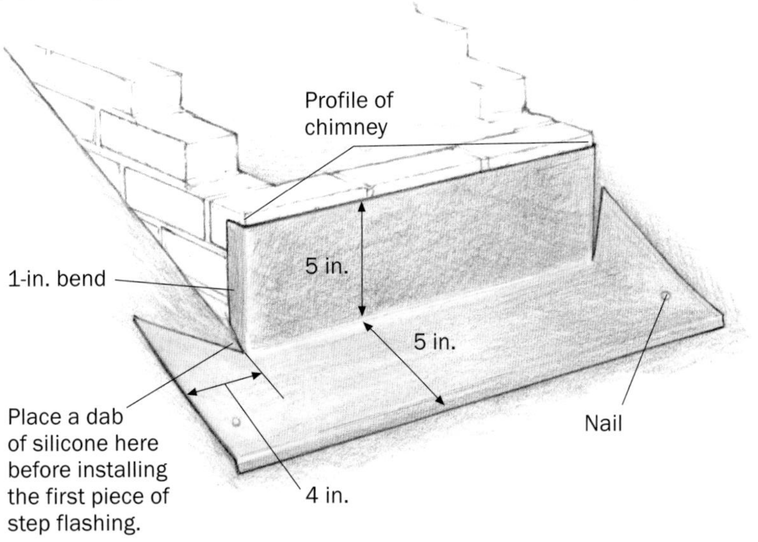

the tabs would grip tightly against the side of the chimney.

To install the apron, I slipped the folded tabs around the chimney and pushed the metal snugly into the corner where the chimney met the roof deck. Then I drove copper roofing nails through the lower half of the apron into the roof deck. On narrow chimneys such as this one, I place the nails only within the 4 in. that extends beyond the sides of the chimney; these nails will be protected by the overlapping shingles. On wide chimneys, I place nails every 24 in. and coat the exposed nails with clear silicone sealant.

Step Flashing Is Woven Into Shingles

Fitting the apron is the most complicated part of the base-flashing process. The rest of this process involves alternating layers of step flashing with shingles (done exactly the same way that you would flash against a sidewall). The right-angle step-flashing cards I have made up for every chimney are 7 in. long and 6 in. wide with the bend creating two 3-in. legs.

To ensure that the vulnerable corners were sealed completely, the first piece of step flashing had to wrap around the corner (see the sidebar on the facing page). After making a crisp, 1-in. bend with my hand seamers, I squeezed a dab of silicone into the corners (see the drawing at left); then I held the flashing tight against the chimney and drove a nail through the outside corner. Following the roofing layout, I overlapped the first piece of step flashing with a shingle. Over the top 7 in. of the shingle, I placed the next piece of step flashing. This I followed with another shingle, then another piece of flashing, and so it went all the way up the side of the chimney. (To keep the roof clean of mortar droppings when I laid up the rest of the chimney, I used scrap shingles to pad the base flashing up to the right height. I would leave these temporary shingles in place until I finished the chimney and cleaned up the mess. Then I would remove the temporary shingles and weave the permanent roof into the flashing.)

I left the topmost piece of the flashing loose until I'd run the step flashing up the other side of the roof. To guarantee a watertight seal at the peak, I trimmed both of the top pieces as shown in the drawing on the facing page. Then I interlocked the two pieces, bent the assembly over the peak and drove a nail through both outside corners.

The chimney flashing that comes up through the peak is symmetrical, so once I finished running the step flashing up this side of the chimney, I repeated the process for all the other sides. If the chimney had come up through the plane of the roof, I would have turned the corner and continued running step flashing up the cricket roof on the backside of the chimney (see the sidebar on p. 48).

Step Flashing Keeps Water on Top of the Roof

The first layer of step flashing seals the corner. To protect the vulnerable corner, the author traces the profile of the chimney on the backside of the first step (photo left). Using tin snips and seamers, he puts a crisp 1-in. bend on the vertical leg (photo right).

Step flashing is woven into the roof. The first step is held tight against the chimney and secured with one nail (photo left). To ensure that rainwater always runs out on the roof, the first step is covered by the next shingle (photo right), which in turn is overlapped by the next layer of step flashing.

GANGING UP ON THE PEAK

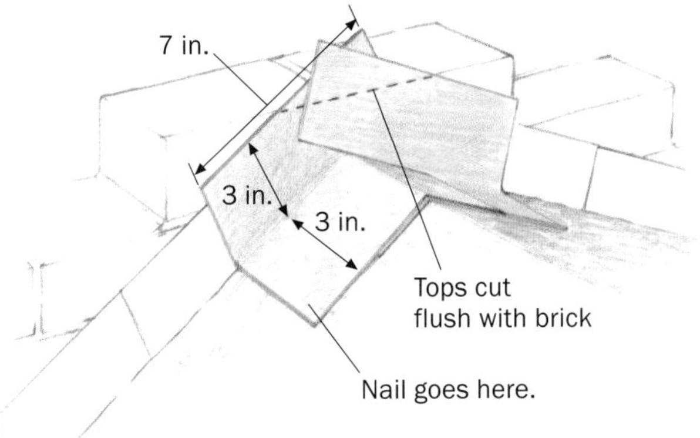

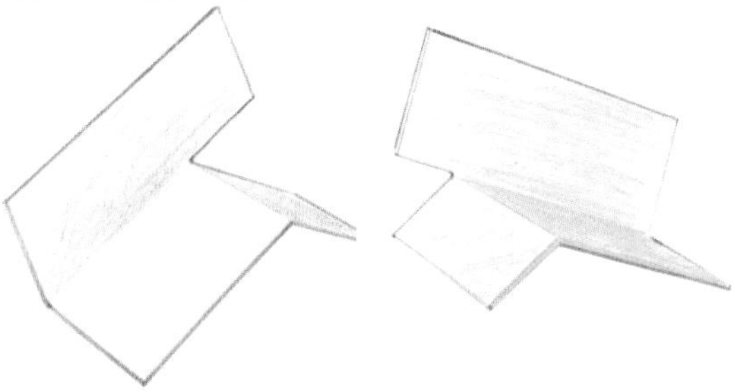

The two pieces of step flashing that meet at the peak are intertwined, folded over the peak and secured with a nail driven through both outside corners.

Building a Cricket

The roofer in me abhors anything that interrupts the flow of water down a roof. So when the back of a chimney faces uphill, I build a cricket to divert water around the edges of the chimney. All I need for this job are two measurements and four pieces of wood. The first measurement is the pitch of the roof, and the second is the width of the chimney. With these measurements in hand, I fabricate the cricket on the ground, then install it as a unit.

After cutting the framing to the dimensions shown in the drawing below, I line up the top of the ridge board with the apex of the gable board and nail the two in the shape of a T. Then I measure and cut two triangular pieces of plywood. Nailing the plywood on my diminutive roof frame completes the cricket.

After nailing the cricket in place behind the chimney, I run shingles over it, weaving the valley into the main roof and, at the same time, installing step flashing against the chimney. Then I'm ready to install the counterflashing.

TWO BOARDS FRAME A CRICKET

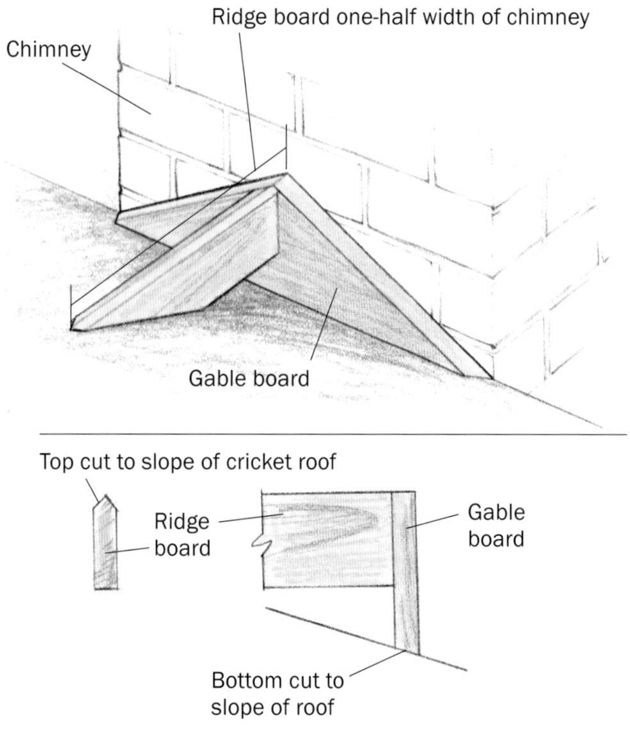

Clamps Hold Counterflashing Until Bricks Are Laid

After all the base flashing was in place, I began setting the counterflashing. One of the advantages of installing the counterflashing when I build the chimney is that it allows me to set the flashing deep into the mortar joint. My counterflashing is bent so that the top lip extends a full 1½ in. into the joint. A ⅜-in. upturned inner lip helps to tie the flashing into the mortar and also serves as a final barrier against water (see the left drawing on the facing page).

As with the base flashing, because this chimney straddled the ridge, I'd made up two front pieces of counterflashing, each of which was 8 in. wider than the chimney. After cutting and fitting each piece to wrap around the chimney, I slipped the flashing over the bricks (see the top photo on the facing page) and secured it with a clamp.

Step Counterflashing Matches the Roofline

With the front piece of counterflashing temporarily clamped in place, I turned to the corners. Each corner piece would cover the first step, with 1 in. wrapping around the front and 1 in. extending beneath the next step up (see the bottom left photo and the right drawing on the facing page). After cutting, fitting and clamping both corner pieces, I set up a bar clamp along the front of the chimney to hold all three pieces of counterflashing. This procedure allowed me to remove the smaller clamps and free up the area above the flashing for when I had to become a bricklayer again.

I used the same basic techniques to cut and fit the remaining steps of counterflashing (see the bottom right photo on the facing page). These pieces do have to be notched to fit into the vertical mortar joints where possible and to lap over the flashing below where

Front Piece Turns the Corners

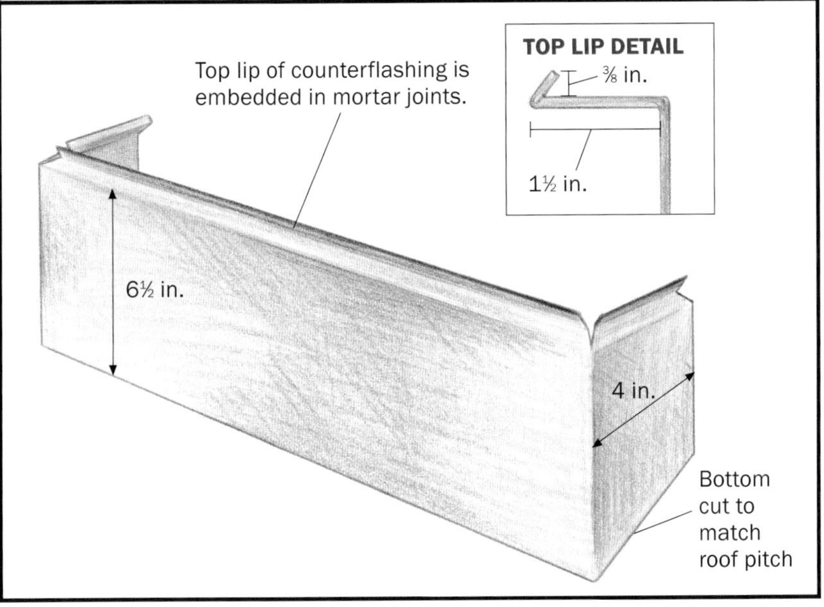

Side Pieces Envelop the Bricks

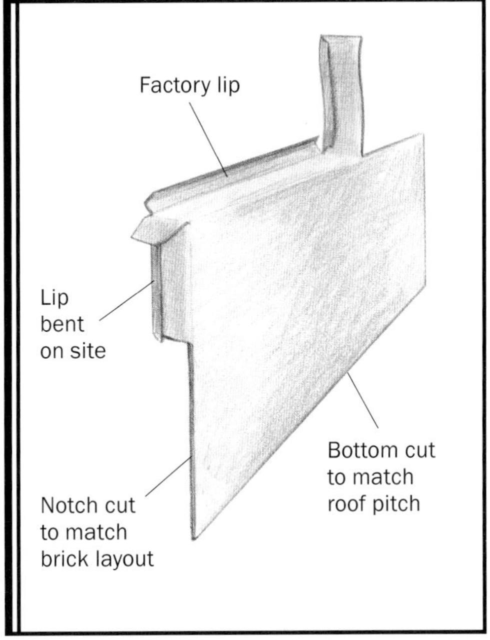

Placing the counterflashing. After bending the sides to match the profile of the chimney, the author slips the counterflashing over the bricks and temporarily secures it with a clamp. Note how the bottom of the side piece is cut to follow the pitch of the roof.

Wrapping the corner. For appearance's sake, the author trims the bottom of the first corner piece to match the roofline and carefully wraps the front corner. To make sure that water never finds an entry point, he cuts the back leg 1½ in. long and folds the top lip up the face of the brick.

Next step up. To secure the flashing without resorting to nails, the author folds the upper levels of counterflashing to lock into the vertical mortar joints (as shown in right drawing above).

Top Piece Straddles the Ridge

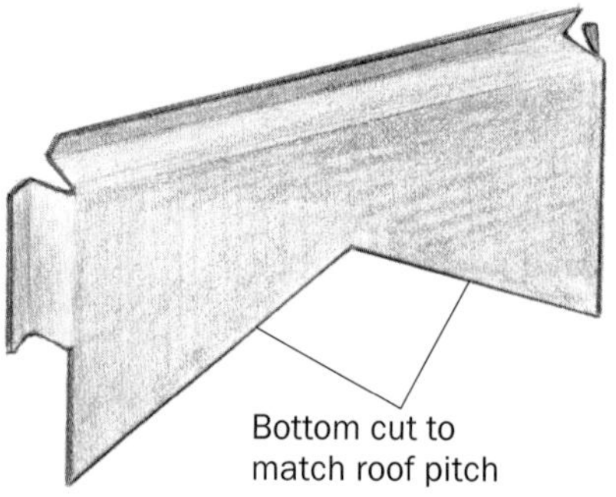

Not the easiest way to lay bricks. Threading bricks, mortar, and trowel through a series of arches takes a steady hand and a bit of patience. Fortunately, the inevitable spills of mortar are easily wiped up with burlap.

impossible. As I fit each piece of counterflashing over the brickwork, I held it in place with its own clamp. To leave enough room to fit the bricks without removing the clamps, I used 24-in. Vise-Grip® (#24SP) C-clamps (American Tool Co.; 847-478-1090). When I reached the two bricks that straddled the peak, I fashioned one piece of flashing with a V-shaped profile on the bottom to cover both sides (see the drawing above).

Embedding the Copper in the Wall

After getting the pieces of counterflashing clamped in place, I mixed up a batch of mortar and began laying bricks above the flashing. I admit that threading a trowel through the jaws of the clamps and then laying the bricks under these temporary arches takes getting used to (see the photo at left). As I pushed each brick into place, mortar that bulged out slid down the flashing. It was messy, but I knew from experience that it would not be difficult to clean up the copper later. I plowed ahead using standard bricklaying techniques, and within a few hours, the flashing was permanently embedded in the structure of the chimney. I was able to remove most of the clamps that day, and I used burlap to clean mortar droppings off the copper.

Keeping Water from Sneaking in From Above

To make sure my brickwork never leaks, I pack all the joints tightly with mortar; I use type-N lime-portland cement mortar, which I'm convinced is more flexible (less prone to hairline cracks) than harder varieties. As I work, I keep an eye on the joints, and when the mortar begins to shrink and pull away from the bricks, I point it with fresh mortar. Then I tool the joint with a jointer that compresses the mortar. To shed water from the top of the chimney, after it has had time to cure, I form and pour a concrete chimney cap in the shape of a hip roof.

***John Carroll** is a builder in Durham, North Carolina, and is the author of two Taunton Press books:* Measuring, Marking & Layout *(1998) and* Working Alone *(1999).*

Preventing Ice Dams

■ BY PAUL FISETTE

The call came on a sunny February afternoon: "The water line to my dishwasher burst, and I can't shut it off. Can you help me out?" I rushed across town to find a former client stuffing towels into the kick space below his kitchen cabinets. I quickly shut off the water to the dishwasher, but nothing happened. Although there wasn't a pipe anywhere near the leak, water still flowed. Puzzled, I made a brief investigation and found the source. The water running down the wall cavity was from an ice dam on the sunlit roof above.

This account is typical of dozens of ice-dam problems I've investigated as a builder, researcher, and consultant. Although individual cases may look different and can result in different types of damage, all ice-dam situations have two things in common: They happen because melting snow pools behind dams of ice at the roof's edge and leaks into the house; and they are avoidable. The symptoms can be treated and the damage repaired, but the key to dealing with ice dams is preventing them in the first place.

Ice Dams Form When Melted Snow Refreezes at Roof Edges

Everyone living in cold climates has seen the sparkling rows of ice that hang like stalactites along eaves. Most people, however, don't stop to understand what causes these ice dams until damage is done. Ice dams need three things to form: snow, heat to melt the snow, and cold to refreeze the melted snow into solid ice (see the photo on p. 51). As little as 1 in. or 2 in. of snow accumulation on the roof can cause ice dams to form. Snow on the upper roof melts, runs under the blanket of snow to the roof's edge and refreezes into a dam of ice, which holds pools of more melted snow. This water eventually backs up under shingles and leaks into the building (see the drawing below).

The cause is no mystery. Heat leaking from living spaces below melts the snow, which trickles down to the colder edge of the roof and refreezes into a dam. Every inch of snow that accumulates on the roof insulates the roof deck a little more, trapping more indoor heat and melting the bottom layer of snow. Frigid outdoor temperatures guarantee a fast and deep freeze at the eaves.

There are a couple of reasons for the loss of heat from living spaces. First, on most homes rafters sit directly on top of exterior walls and leave little room for insulation between the top of the wall and the underside of the roof sheathing. Second, some builders aren't particularly fussy when it comes to stopping the movement of warm indoor air into this critical area.

Avoid Expensive Damage by Recognizing the Signs of Ice Damming

Ice dams cause millions of dollars in damage every year. Much of the damage is apparent. We easily recognize water-stained ceilings; dislodged roof shingles; sagging, ice-filled

How Ice Dams Form

Hot air escapes from living spaces, melting snow on the roof, which trickles to the colder edge of the roof where it refreezes into a dam. Every inch of snow that accumulates on the roof insulates the roof deck a little more, trapping more indoor heat and melting the bottom layer of snow.

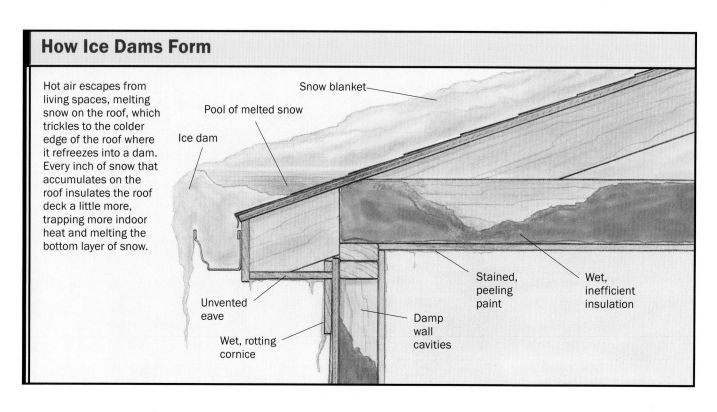

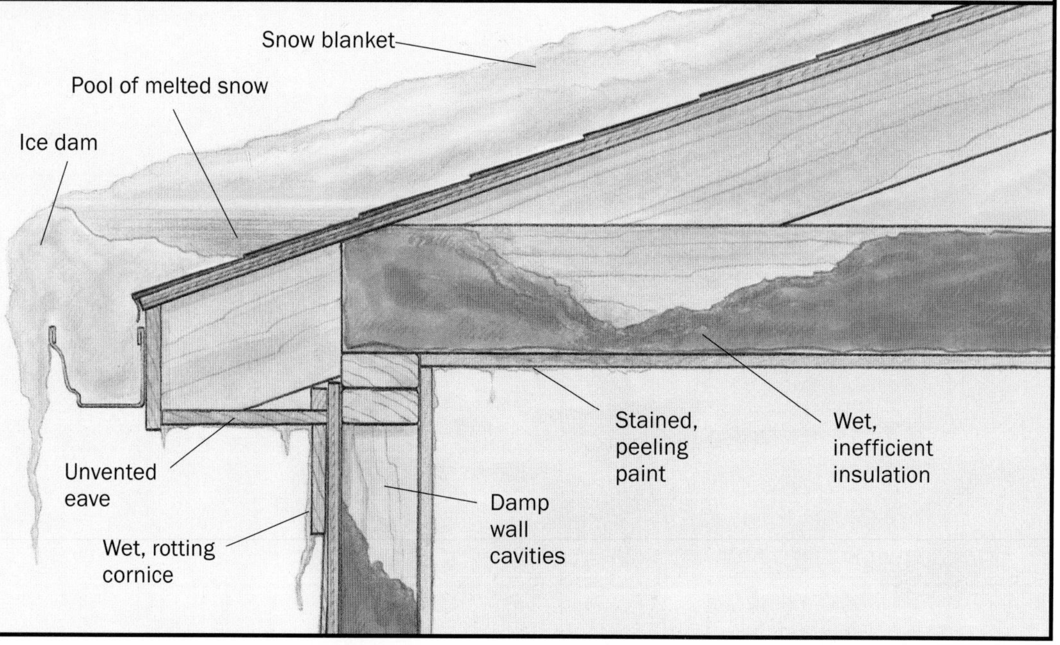

gutters; peeling paint; and damaged plaster. So check your home carefully when you notice any of these signs.

Not all damage is as obvious as water stains on the ceiling, however, and some hidden damage can go unchecked. Insulation is one of the biggest hidden victims of leaks. Roof leaks dampen attic insulation, which in the short term loses some of its insulating ability. Over the long term, water-soaked insulation compresses so that, even after it dries, the insulation isn't as thick. Thinner insulation means lower R-values. As more heat leaks from living areas into the attic, it's more likely that ice dams will form and cause leaks. The more water that leaks through the roof, the wetter and more compressed the insulation becomes. Cellulose insulation is particularly vulnerable here. It's a dangerous cycle, and as a result, you pay more to heat and cool your house.

There's more: In the wall, water leaks soak the top layer of insulation and cause it to sag, leaving uninsulated voids at the top of the wall (see the drawing on the facing page). More heated air escapes. More important, moisture in the wall gets trapped between the exterior plywood sheathing and interior vapor barrier. The result is smelly, rotting wall cavities. Structural framing members can decay; metal fasteners can corrode; mold and mildew can form on wall surfaces as a result of elevated humidity levels; and exterior and interior paint can blister and peel.

Peeling wall paint deserves special attention because its cause may be difficult to recognize. It's unlikely that interior or exterior wall paint will blister or peel while ice dams are present. Paint peels long after ice and all signs of a roof leak have evaporated. The message is simple. Investigate even when there doesn't appear to be a leak. Look at the underside of the roof sheathing and roof trim to make sure they're not wet. Check insulation for dampness.

It's often difficult to follow the path of water that penetrates a roof. However, patching the roof leak won't solve the problem. You do need to make sure the roof sheathing hasn't rotted or that other, less obvious problems in your ceiling or walls haven't developed. Once you've got a handle on the damage, it's good to detail a comprehensive plan to fix the damage, but first you need to solve the problem.

Keep the Whole Roof Cold to Avoid Ice Dams

Damage by ice dams can be prevented in two ways: Maintain the entire roof surface at ambient outdoor temperatures so that an ice dam never forms; or build a roof that won't leak if an ice dam does form (see the drawing on p. 54).

The first choice is definitely the best. Cold roofs make sense because they make the cold outside air work for you. If you keep the roof as cold as the outdoor air, you solve the problem. Look at the roofs of unheated sheds. Ice dams don't form on them because the air inside the roof is as cold as the air outside.

It's relatively easy to keep a roof cold in new construction: Design the house to include plenty of ceiling insulation and effective roof ventilation, and make sure heat doesn't escape from the house into the attic. Insulation retards the flow of heat from the heated interior to the roof surface; good ventilation keeps the roof sheathing cold; well-sealed walls and ceilings keep the heat where it belongs.

In an existing house, this approach may be more difficult because often you're stuck with less than desirable conditions. This opportunity is a good point to take a closer look at the issues that will guide your strategy.

How Ice Dams Are Prevented

Cold outside air flowing from soffit to ridge along the underside of roof sheathing keeps the roof cold. Good insulation and well-sealed interior spaces keep heat inside. A baffle above the exterior wall will help protect attic insulation from blasts of cold air.

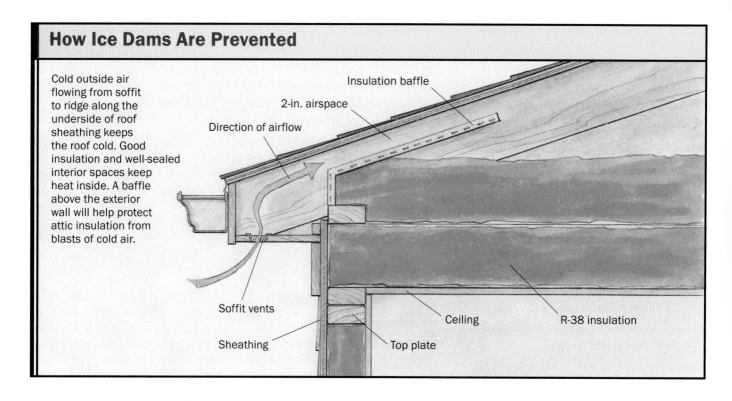

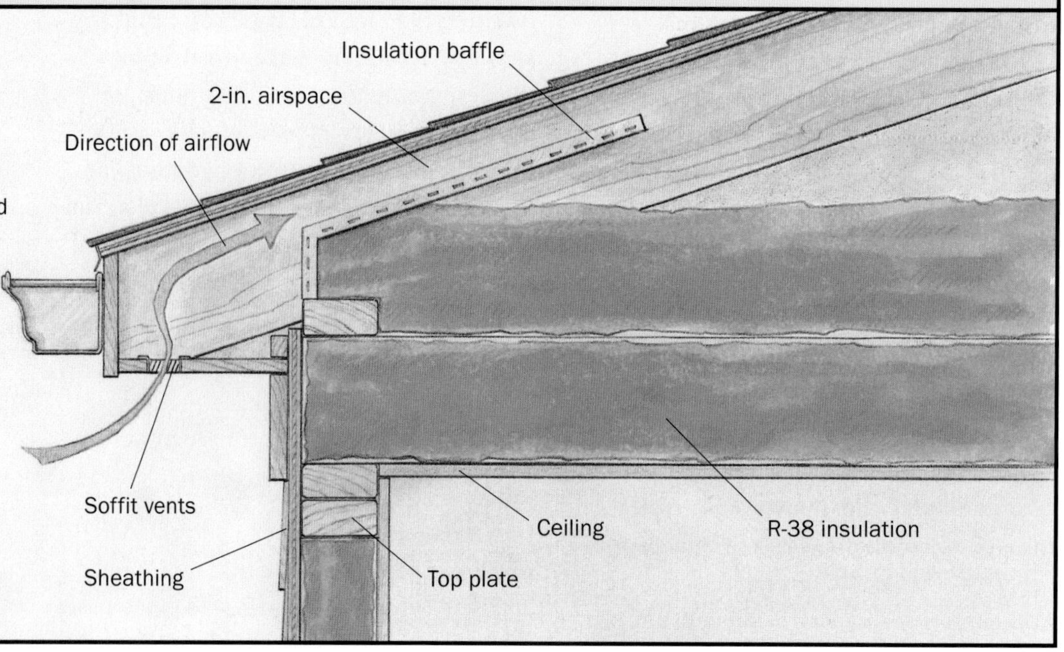

Ice dams are created by heat lost from the house. So when possible, develop a strategy that includes plugging all heat leaks into unheated spaces.

When Treating Symptoms Is the Only Choice

The list of efforts to deal with ice dams is long. The problem I have with most of these solutions is that they treat the symptoms of ice damming and don't deal with the root cause, heat loss.

For instance, some people assume they can fix the problem by installing a metal roof. Metal roofs are common in snow country, so they must work, right? Well, a deeply pitched metal roof does, in a sense, thumb its nose at ice dams. Metal roofs are slippery enough to shed snow before it causes ice problems. However, metal roofs are expensive, and they do not substitute for adequate insulation.

Or you might consider using sheet-metal ice belts if you don't mind the look of a shiny 2-ft.-wide metal strip strung along the edge of your roof. Ice belts are reasonable choices for some patch-and-fix jobs on existing houses. This eave-flashing system tries to do what metal roofing does, which is shed snow and ice before they cause problems. Unfortunately, it doesn't always work. Often, a secondary ice dam develops on the roof just above the top edge of the metal strip, so the problem simply moves from one part of the roof to another. Ice belts are sold in 32-in. by 36-in. pieces and come with fastening hardware for about $12* per panel.

Many people install self-sticking rubberized sheets under roof shingles wherever ponding of water against an ice dam is possible: along the eaves, around the chimneys, in valleys, around skylights, and around vent stacks. The theory is, if water leaks through the shingles, the waterproof underlayment will provide a second line of defense.

These products are sold in 3-ft. by 75-ft. rolls for about $80 per roll. They adhere directly to clean roof decking. Roof shingles are nailed to the deck through the membrane, which is self-healing and seals nail penetrations automatically. Grace Ice & Water Shield® (W. R. Grace Co.; 617-876-1400; www.na.graceconstruction.com) and Ice and Water Barrier (Bird Roofing Products

Inc., 1077 Pleasant St., Norwood, MA 02062; 800-247-3462) are two common brands. Installing such products is a reasonable alternative for many existing structures where real cures either are not possible or cost effective.

Heat tape is another favorite solution to ice damming. I have never seen a zigzag arrangement of electrically heated cable solve an ice-dam problem. Electricity heats the cable, so you throw more costly energy at the problem (keep in mind that ice dams are a heat-loss problem). Heat tape is expensive to install and to use. Over time, the tape makes shingles brittle and creates a fire risk, and its loose fasteners allow water to leak into the roof. Take a good look at roofs equipped with heat tape. The electric cable creates an ice dam just above it. My advice is don't waste your time or money here.

Shoveling snow and chipping ice from the edge of a roof is my least favorite of all solutions. People attack mounds of snow and roof ice with hammers, shovels, picks, snow rakes, crowbars, and, new to my list, chainsaws. The theory is obvious: Where there's no snow or ice, there's no leaking water. Some people have even carved channels in the ice to let trapped water flow out.

Whatever plan you decide to follow, focus on the cause. Ice dams are created by heat lost from the house. So whenever possible, develop a strategy that includes plugging all heat leaks into unheated spaces. You can use urethane spray foam in a can, caulking, packed cellulose, or weatherstripping to seal leaks made by wiring, plumbing, attic hatches, chimneys, interior partition walls, and bathroom exhaust fans.

There are no excuses for ice-dam problems in new construction. But in existing houses, you can improve ventilation, upgrade insulation, and block as many air leaks as you can. However, cures for existing structures are often elusive and expensive, and in some cases you may have to settle for merely treating the symptoms. The payback is damage prevented.

Prevention Is the Key in New Construction

Houses in heating climates should be equipped with ceiling insulation of at least R-38, which equals about 12 in. of fiberglass or cellulose. The ceiling insulation should be of continuous and consistent depth.

As I mentioned earlier, the biggest problem area is just above the exterior wall. Raised-heel trusses or roof-framing details that allow for R-38 above the exterior wall—while maintaining room for airflow from the eave to the ridge—should be used in new construction (see the photo below). (In existing structures, where there's little space between the top plate and the underside of the roof sheathing, install R-6-per-in. insulating foam). Insulation slows conductive heat loss, but an effort must be made to block the flow of warm indoor air into the attic or roof. Even small holes allow significant volumes of warm indoor air to pass into the attic. In new construction, it's best to avoid ceiling penetrations (such as recessed lights) whenever possible.

Elevated rafters make room for insulation. The builder made room for a thick blanket of insulation above the exterior wall by nailing the heels of these rafters to a toe board that sits on top of the ceiling joists. This construction method also maintains an adequate ventilation space between the sheathing and the insulation so that air can flow from the soffit vent to the ridge vent.

Sources

ADO Products
21800 129th Ave. N.
Rogers, MN 55374
(866) 240-4933
Sells high-impact polystyrene chutes.

Insul-Mart Inc.
775 Fall River Ave.
Seekonk, MA 02771-5627
(401) 831-0800
Sells polystyrene chutes.

Insul-Tray Inc.
E. 1881 Crestview Dr.
Shelton, WA 98584
(206) 427-5930
Manufactures water-resistant corrugated-cardboard baffles.

Three Ways to Vent a Cathedral Ceiling

Ventilation in cathedral ceilings is a little trickier than in conventional ceilings, but it is just as necessary to prevent ice damming. Careful sealing of heated spaces and good insulation and ventilation all are required.

PLASTIC CHUTES CREATE AIRSPACE OVER FIBERGLASS BATTS

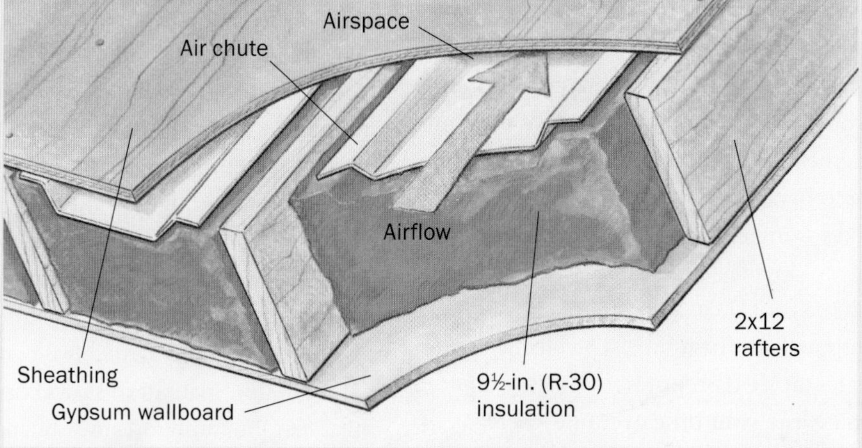

WIDE I-BEAM RAFTERS ALLOW EXTRA INSULATION AND A GENEROUS AIRSPACE

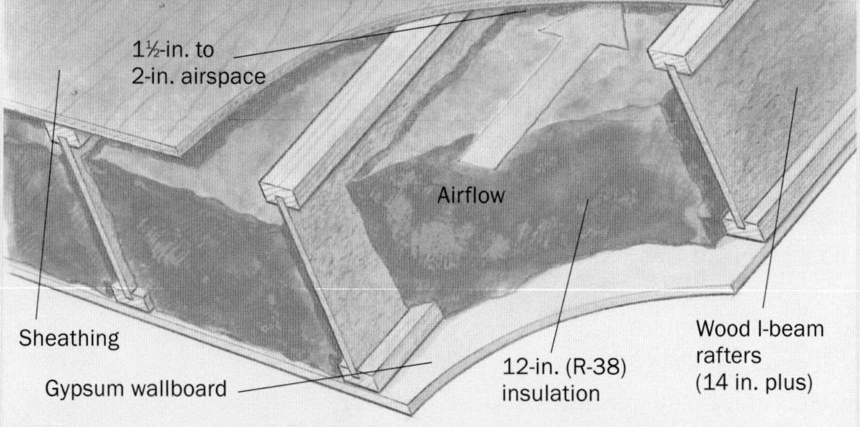

FOAM BOARD AND FURRING STRIPS CREATE AN AIRSPACE ABOVE 10-IN. RAFTERS

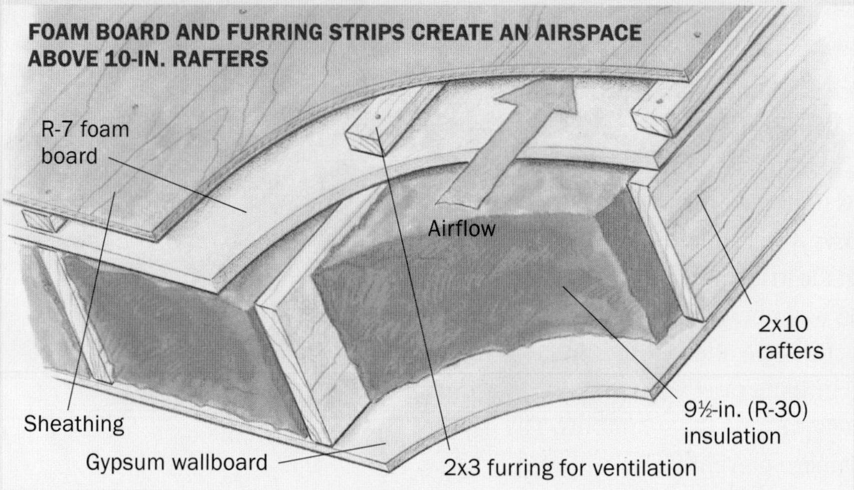

Continuous vents keep roof washed in cold air. A ridge vent draws cold air from the continuous soffit vents uniformly across the underside of the roof sheathing.

Soffit-to-ridge ventilation is the most effective way to cool roof sheathing. Power vents, turbines, roof vents and gable louvers don't work as well. Soffit and ridge vents should run continuously along the length of the house (see the photo above). A baffled ridge vent is best because it exhausts attic air regardless of wind direction. Exhaust pressure created by the ridge vent sucks cold air into the attic through the soffit vents.

In cathedral ceilings, it's important to provide a 2-in. space or air chute between the top of the insulation and the underside of the roof sheathing. The incoming air washes the underside of the roof sheathing with a continuous flow of cold air. The construction of cathedral ceilings requires some special consideration because the ceiling and the roof are the same structure (see the sidebar on the facing page).

I want to add a caution here. Cold air flowing in through soffit vents in any kind of roof can blow loose-fill fiberglass or cellulose insulation out of the way. It will find pathways through batt or roll fiberglass. Unless the pitch of the roof prohibits it, install cardboard or polystyrene baffles in the attic space above exterior walls (see the drawing on p. 54) to protect insulation from cold air.

**Prices noted are from 1995.*

***Paul Fisette** is Director, Associate Professor, of the Building Materials and Wood Technology program at the University of Massachusetts in Amherst.*

Four Ways to Shingle a Valley

■ BY MIKE GUERTIN

A lot of today's new-home designs include multiple gables and roof configurations. When two roof planes meet at an inside corner, a valley is created. Because valleys collect and channel a greater volume of water than a single roof plane, I always make an extra effort to design and build them as watertight as possible.

For this article, I mocked up a section of roof to show four ways to shingle valleys. The mock-up allows me to show all four methods in a similar context. I used three-tab shingles for two of the methods and laminated shingles for the other two. When shingling a roof with three-tab shingles, it's easiest to shingle the roof planes first, working toward the valley. Laminated shingles let you start in the valley and work outward, an advantage in some situations.

Preparing a Roof: All Valleys Start the Same

Regardless of the shingling method, every successful valley installation begins with proper roof preparation. Taking the right steps before applying the shingles not only goes a long way toward preventing roof leaks, but it also helps to cushion the shingles (or metal valley) against the ragged edges of the roof sheathing at the centerline of the valley.

In the past, I've used several different methods to prepare a valley, including lining the valley with aluminum-coil flashing or roll roofing, and even cementing together layers of #30 felt paper with asphalt roof cement (a messy job). Today, fortunately, we have a simpler and more effective material at our disposal: waterproof shingle underlayment (WSU). Examples are Grace Ice & Water Shield (W. R. Grace; www.na.grace-construction.com), WeatherWatch® (GAF; www.gaf.com), and Moisture Wrap (Tamko®; www.tamko.com). These peel-and-stick membranes seal around nails and are pretty easy to work with. A release sheet on the back keeps the membrane from adhering until you remove it.

I begin the prep work by sweeping off any sawdust or other debris. I also set any sheathing nails that stick up from the roof plane and could puncture or wear through the WSU or shingles. As with any asphalt-shingling job, I install a 9-in. to 12-in. strip of WSU along the eaves' edges, and I cross-lap the strips at the valley. These strips go under the drip edge.

I overlap the inside corner of the drip edge at the valley to minimize sharp edges that could cut into the WSU over time. Because I build in snow country and along the coast, I include extra protection along the eaves. After rolling out full 3-ft. widths of WSU over the drip edge and up the roof sheathing, I again overlap the WSU at the valley (see the top photo at right).

For the valley itself, I snap a chalkline parallel to the valley at a distance half the width of the WSU (see the center photo at right). I cut a sheet of WSU about 3 ft. longer than the overall length of the valley and roll it into the valley with the release sheet still on. I line up one edge of the WSU with my snapped line and staple down that edge of the sheet every couple of feet (see the bottom photo at right). Next, I fold back the loose edge, remove the release sheet from that half of the WSU (see the top left photo on p. 60) and then roll the sticky side down to the roof sheathing (see the top right photo on p. 60). In colder weather, it may be necessary to nail down the sheet if it doesn't stick immediately.

Waterproof shingle underlayment (WSU) protects the eaves. After the roof is swept clear of debris and the drip edge is installed, a full width of WSU is unrolled along the edge of the eaves, overlapping at the valley.

A guide line helps to keep the WSU straight. Before the WSU is rolled out, a chalkline is snapped parallel to the valley (above). The sheet then is tacked along the line to keep it aligned with the valley during installation (left).

Peel and stick, one side at a time. After the WSU is tacked in place, the free side is folded back, and the release sheet is peeled off that half (far right). The sticky side then is rolled out onto the sheathing (right). Repeat the process for the second side, taking care the WSU is adhered fully to the center of the valley (below).

I tug the stapled edge free, fold it on top of the stuck side, and remove its release sheet. I then roll the sheet back down onto the roof sheathing (see the bottom photo above), taking extra care to keep the sheet tight in the valley center without forming hollows. Although WSU can stretch a bit, any hollow left beneath the sheet is likely to tear when stepped on. After the sheet is fully adhered to the valley, I trim the excess material along the drip edge. Then, as with any shingled roof, I finish the prep work by nailing down #15 felt paper and snapping horizontal chalklines for the shingle courses.

Woven Valley

Woven valleys seem to have fallen out of favor in most regions of the country, mainly because they are the slowest to install. But woven valleys are the most weather-resistant, and unlike other valleys, they don't require sealing with messy roofing cement.

You can weave a valley working either into or out of it. However, working out of the valley requires laminated shingles, the multiple layers of which can make a woven valley look very bulky. On the other hand, working three-tab shingles out of the valley would make aligning their slots almost impossible. For the purposes of this article, I'll describe the weaving process working into the valley with three-tab shingles.

I start by shingling both roof planes to within about a half shingle of the valley center, leaving the ends staggered. Then the starter shingles and the first course are nailed in, with the shingle from the larger roof plane lapping over the shingle from the smaller plane on each course (see the top left photo on p. 62). (Every valley method, except the open metal valley, begins with a woven course.)

Method Pros and Cons

Woven

The shingles from both roof planes overlap on each course.

ADVANTAGES

- Shingles don't need to be cut.
- Shingles don't rely on asphalt roof cement to be sealed.
- Interlocking weave provides double coverage, which makes it the most weather- and wind-resistant choice.

DISADVANTAGES

- Both sides must be shingled at the same time, which slows installation.
- Hollows under the weave can be punctured when stepped on.
- Hollows in the weave make the valley look uneven from the ground.
- There is a greater potential for splitting from thermal changes.
- Extra-thick laminated shingles may make the valley look bulky.

Open Metal

Metal liner or flashing lines the valley and is left exposed.

ADVANTAGES

- Variety of metals can be used.
- Metal valley creates the most decorative appearance.
- Method works well with laminated shingles (no bulky look or telegraphing).
- Metal valley is durable.

DISADVANTAGES

- Metal valley has to be prefabricated.
- This method relies on proper shingle overlap and roof cement to keep out water.
- Nailing through metal must be avoided.
- Metal valleys tend to be more expensive.
- Metal valleys require cutting shingle edges and dubbing (see the sidebar on p. 64) shingle corners from both roof planes.

Closed-Cut

Shingles from one roof plane cross the valley and are lapped by shingles from the other side, which are cut in a line along the valley.

ADVANTAGES

- The two converging roof planes don't have to be shingled at the same time.
- Cut valleys are usually faster than weaving.
- Cut valleys provide a crisp look.

DISADVANTAGES

- Cut valleys provide single coverage through the center of the valley.
- Shingles from one side have to be cut back and dubbed.
- Cut edge must be sealed with roof cement.

New: Long Island Valleys

This new method starts out like a cut valley, but instead of lapping and cutting the shingles on the second side, you form the "cut" line with a row of shingles turned on edge that run up the valley.

ADVANTAGES

- This method looks like a cut valley without cutting shingle edges or dubbing corners.
- This technique is faster than any other valley-shingling method.

DISADVANTAGES

- This method works only with laminated shingles.
- Long Island valley system is not yet approved by shingle manufacturers.

I keep all nails at least 6 in. away from the valley center. A quick way to gauge this distance is with the span between my thumb and index finger (see the top right photo below). I continue up the valley, overlapping shingles on each course and pressing each shingle into the valley center to halt bridging.

As the shingle courses march up the valley, it's necessary to insert either single- or double-tab shingles to make sure the valley shingle wraps through far enough (see the bottom left photo below). The minimum distance I let a shingle lap across the valley (see the bottom right photo below) is 8 in. from the valley center to the outer edge of the shingle's nailing strip (or the distance from my thumb to the end of my middle finger).

As I press each shingle into the valley, I usually put an extra nail in the shingle corner farthest from the valley center to help hold the shingle flat. I also run full shingles through the valley to keep end joints between shingles as far away from the valley as possible.

Shingle edges colored
for photo purposes.

The first course is always a weave. For every valley variation, except the open metal valley, the first course is a weave. The shingle from the larger roof plane overlaps the shingle from the smaller plane.

A handy nailing gauge. Nails always should be kept at least 6 in., or the distance between an outstretched thumb and index finger, away from the center of the valley.

Insert a tab to stretch the course. Every course should wrap through the valley with a full shingle. A single- or a double-tab shingle may be installed to allow the shingle to extend through the valley (left) a minimum distance of 8 in., or the distance from thumb to middle finger (above).

Open Metal Valley

I consider open metal valleys to be the most decorative of the bunch. They're used traditionally with wood shingles, tile, or slate roofing, but they also work well with asphalt shingles, especially today's heavy laminated shingles, which can be difficult to bend tightly into the valley. Open valleys are most durable when lined with copper; lead-coated copper; or enameled, galvanized, or stainless steel. Aluminum also can be used, but the uncoated mill-finish aluminum commonly sold won't last as long as heavy-gauge color-coated aluminum.

With a metal valley, the first hurdle is having the metal liner fabricated. My HVAC duct fabricator custom-bends all the pieces for my projects. The liner should be 2 ft. to 3 ft. wide, which leaves 12 in. to 18 in. on each side of the valley. Also, each liner panel should be no more than 8 ft. long to allow for lengthwise expansion. I like to have a bent, inverted V, about 1 in. high, down the middle of each panel (see the top photo at right). The V helps to resist the flow of water across the valley by channeling it down the center. The crimp also stiffens the liner lengthwise and adds a flex point for widthwise expansion, which helps to keep the liner from wrinkling on hot days.

The metal liner panels must be fastened to allow for expansion and contraction. The simplest method is to trap the edges of the liner with the heads of nails driven every 12 in. to 16 in. I butt the nail shank to the edge of the metal and drive the head until it just touches the metal without dimpling it (see the bottom right photo at right). An even better method uses clips that interlock a hem along both edges of the liner panel (see the bottom left photo at right). Just make sure the clips or nails that are snug against the panels are made of the same metal to avoid a galvanic reaction. Also, upper liner panels should overlap lower ones by 6 in. to 8 in. as they progress up the valley.

An extra crimp for extra protection. An inverted V in the center of the valley liner helps to channel water down the valley while allowing for expansion and contraction widthwise.

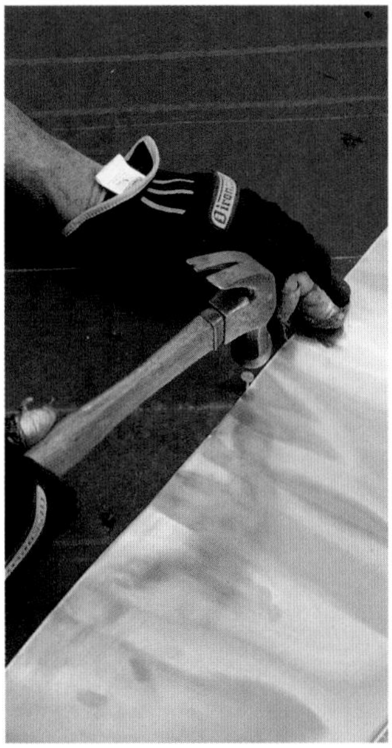

Two ways of holding down the liner. To allow the liner to expand lengthwise, fasteners cannot be driven through the liner. Instead, nails can be driven snug along the edge of the liner (right), or simple clips can engage a hem along the edge (above).

Shingles should overlap the edge of the liner panels by at least 6 in., and at least 3 in. of metal should be exposed on each side of the valley center for appearance and water flow. Shingles can be trimmed and dubbed (see the sidebar on p. 64) before they're nailed in, or run long across the valley and trimmed and dubbed after.

Dubbing Shingle Corners

When water rushes down a valley, it has to be channeled downhill as it passes every shingle course. If the upper corners of the shingles are left straight in an open metal valley or a cut valley, water can be diverted away from the valley, causing leaks to appear several feet from the valley. To help prevent these leaks from occurring, the upper corner of each valley shingle should be clipped or dubbed.

To dub a shingle, measure approximately 2 in. down the cut edge of the shingle (above and right) and make a square cut back to the top edge of the shingle. This removes a small triangle of shingle. Any water that hits the dubbed corner of the shingle should now be diverted safely down the valley and away from the main part of the roof.

Dubbing corners can prevent leaks. Dubbing or clipping the corners of shingles along the valley keeps water flowing downhill along a cut edge. Corners can be clipped either before the shingles are installed or in place before the edge is cemented.

For the first method, I begin by snapping chalklines on the liner 3 in. from the center to guide shingle placement. Working toward the valley as I did here, I measure, cut, and dub the shingles for a half-dozen or so courses. I then spread a 2-in. to 3-in. band of roof cement up the valley (see the top left photo on the facing page). As I nail in each course, I run the band of roof cement on top of the shingle down to the adhesion strip (see the top right photo on the facing page). For an open metal valley, keep all shingle nails at least ½ in. outside the edge of the liner.

For the second method, I overlap the valley with each course of shingles (see the left center photo on the facing page). Next, I snap a chalkline on my desired cutline. With a shield of sheet metal or a shingle slipped in between the shingles and metal liner, I start at the top and cut the shingles with a hook-blade knife (see the right center photo on the facing page). The trick to cutting a smooth, straight line is to cut through one shingle at a time. I then go back and dub the top corner of each shingle.

The cut edges of shingles need to be bedded in a double ribbon of asphalt roof cement to resist water and to bond them to the valley liner. The quickest method is to use a caulking gun, running one bead with the nozzle all the way under the shingles (see the bottom left photo on the facing page) and a second bead with my finger on the end of the gun as a spacer (see the bottom center photo on the facing page). Finally, double beads have to be continued down onto the top laps of each shingle (see the bottom right photo on the facing page).

Start spreadin' the goo. If you cut the shingles before nailing them in, spread a band of roof cement on the metal liner first (far left). As each course goes on, the roof cement is spread on the top lap of each shingle as well (left).

Nail now, cut later. A second way to install the shingles in an open valley is to run them long through the valley (far left). Then, after snapping a cutline, trim the shingles with a hook blade in a utility knife, using sheet metal to protect the valley liner (left).

A controlled double bead. For a quick and easy double bead of sealer, run the first bead with the nozzle of the gun inserted all the way under the shingles (left). A finger on the nozzle spaces the second bead (center), and the beads continue on the top lap of each shingle to complete the seal (right).

No dubbing needed here. Working out of the valley with the closed-cut method, snap the cutline (left), then step the shingle back from the line to eliminate the need for dubbing the corners. Here, the gauge shingle marks the guide line (right).

Snap and cut. When all the shingles are installed to the guide line, the cutline is resnapped (left). With sheet metal protecting the lower layer of shingles, the shingles on top are cut back to the line (right). The cut edge then is sealed with roof cement.

Closed-Cut Valley

Even if I work toward the valley with three-tab shingles, cut valleys are faster than woven valleys hands down. With a closed-cut valley, there's no need to shingle the two roof planes at the same time. Plus, cutting the closed-cut valley shingles can happen after the rest of the roof is shingled.

I begin the second side (the side that will be cut) by snapping a cutline 2 in. to 3 in. from the valley center (see the photos at left). Keeping the cutline away from the center of the valley creates a better watercourse for runoff and tends to hide discrepancies in the line after the shingles are cut.

Here's one of the big advantages of working out of the valley with laminated shingles. To establish a line to guide the placement of the shingles on the second roof plane, I place the lowest shingle on the course line so that the cutline meets the shingle 2 in. down from the top edge. In this position, there is no need to dub the corners of the shingles. As I did on the other side, I mark the location of the outside corner of the shingle on the felt, repeat the process at the top of the valley, and snap a chalkline. I then install the shingles, lining the top edge with the course line and the outside corner with the guide line, letting the other edges run through the valley.

When the shingles are all in, I resnap my cutline. For protection, I insert a metal sheet between the shingles from both roof planes, and I trim the shingles one at a time with a hook blade in my utility knife. I finish the cut edge with a double bead of roof cement, the same as for the open metal valley.

Long Island Valley

A young roofer from Long Island first showed me what I now call a Long Island valley. This valley looks the same as a cut valley, only it's faster to install. Although this valley system isn't entirely new, you won't find it described on shingle wrappers.

Closed-Cut and Long Island Valleys Begin the Same

For the next two shingling methods, closed-cut and Long Island valleys, I work out of the valley using laminated shingles instead of working toward the valley with three-tab shingles. Shingling both a closed-cut valley and a Long Island valley is the same for the first half of the process. Working on the smaller roof plane, I set a shingle on the first course line. I place the shingle so that one edge is 2 ft. away from the valley center at the nail line. I mark the shingle where the valley center crosses the top edge, and I mark the roof at the top outside corner of the shingle (see the top photo at right).

I then move the shingle to the uppermost course on the roof plane, line up the mark on the shingle with the valley center and again mark the corner (see the center photo at right).

A chalkline snapped between this mark and the lower mark forms my guide line.

I always weave my first course. But after that I run the shingles up the valley, aligning the top edges with the course lines and the top outside corners with the guide line (see the bottom photo at right). I nail the shingles normally, except that I keep nails at least 6 in. away from the valley center. Now I'm ready for the other roof plane.

To shingle out of the valley with laminated shingles, lay a shingle down as a gauge and mark the corner at the bottom (above) and top (below) of the roof. After snapping the guide line between marks, install the shingles with one corner on the line (bottom). Snap a chalkline 3 in. from the valley, and you're ready to complete the valley.

A row of shingles creates the valley. For a Long Island valley, a row of shingles is bedded in roof cement (right) and installed with their top edge on what would be the cutline (far right).

No guide line needed. Shingles go in with one corner butted to the valley shingle.

Minimal sealing. A small dab of roof cement seals each shingle corner.

However, it seems to be a viable weather-resistant method. The only drawback is that Long Island valleys work only with laminated, random-pattern shingles. This method cannot be used with three-tab shingles.

When I've finished shingling the first roof plane, I snap a chalkline 2 in to 3 in. away from the valley center, just as I did with the closed-cut valley. Next, I smear roof cement a couple of inches away from my snapped line (see the top photos above). Then I install a line of shingles up the valley with the top edge aligned with my snapped line. The lowest of these valley shingles is cut back at an angle in line with the lowest course line.

I install the shingles for each course with the lower corner lined up with the edge of the valley shingles. The result of this layout leaves a small triangle of the valley shingle that is exposed on each course. From the ground, the Long Island valley is indistinguishable from a cut valley. The sealing is easier, too: Just a half-dollar-size dab of roof cement under each corner where it laps over the valley shingle is all that is required.

***Mike Guertin**, author of* Roofing with Asphalt Shingles *(The Taunton Press, 2002), is a builder, construction consultant and contributing editor to* Fine Homebuilding.

Laying Out Three-Tab Shingles

■ BY JOHN CARROLL

Many roofers take pride in the fact that they can shingle a house without the benefit of measured lines. It can't be denied that such people install leak-proof roofs that look pretty good from the ground. Unfortunately, their eyeballed roofs often have wavy, inconsistent courses, and when viewed from atop the house, they look simply unprofessional. When I finish a roof, I enjoy looking at straight courses, and I don't begrudge myself the half-hour or so it took to measure and strike lines. More than that, I'm convinced that I recover the time invested in laying out the roof as I nail down shingle after shingle without worrying about the courses getting wavy or crooked.

For the sake of simplicity, I'll limit my discussion to the ubiquitous three-tab asphalt roof shingle, scorned by aesthetes but found on houses from the Carolinas to California.

On a rectangular roof without dormers, valleys, or other obstructions, there are three basic layout steps: establishing the overhang, striking the bond lines, and striking the horizontal lines.

Establishing the Shingle Overhang

Before shingling a roof, it is essential to know how far the shingles will overhang the bottom (or eaves) and sides (or rakes) of the roof deck. Ideally, all trim has been installed along the roof edges, and if used, metal drip edge is also in place. In these cases I leave a 1-in. overhang along the eaves and the rakes of the roof (see the drawing on pp. 70–71). Most shingle manufacturers recommend a ¼-in. to ⅜-in. overhang, presumably to reduce the chance of the wind snagging the edge of the roof.

Unfortunately, eaves and rakes (especially those on older houses) often diverge more than ⅜ in. from a straight line. To compensate for irregularities in the straightness of rake boards and fascia boards, I've found that a 1-in. overhang allows me to work proficiently and provides for a straight, secure roof. I've never had a problem with shingles blowing off.

If I install the shingles before the roof trim is complete, I need to know how far the

Basics of Asphalt-Roofing Layout

The author overhangs the asphalt shingles 1 in. on both the rake and the eaves. Before nailing down any shingles, he strikes two vertical bond lines and a series of horizontal lines to ensure that the tabs and the courses will line up neatly. After the layout is finished, nailing on the shingles is a breeze.

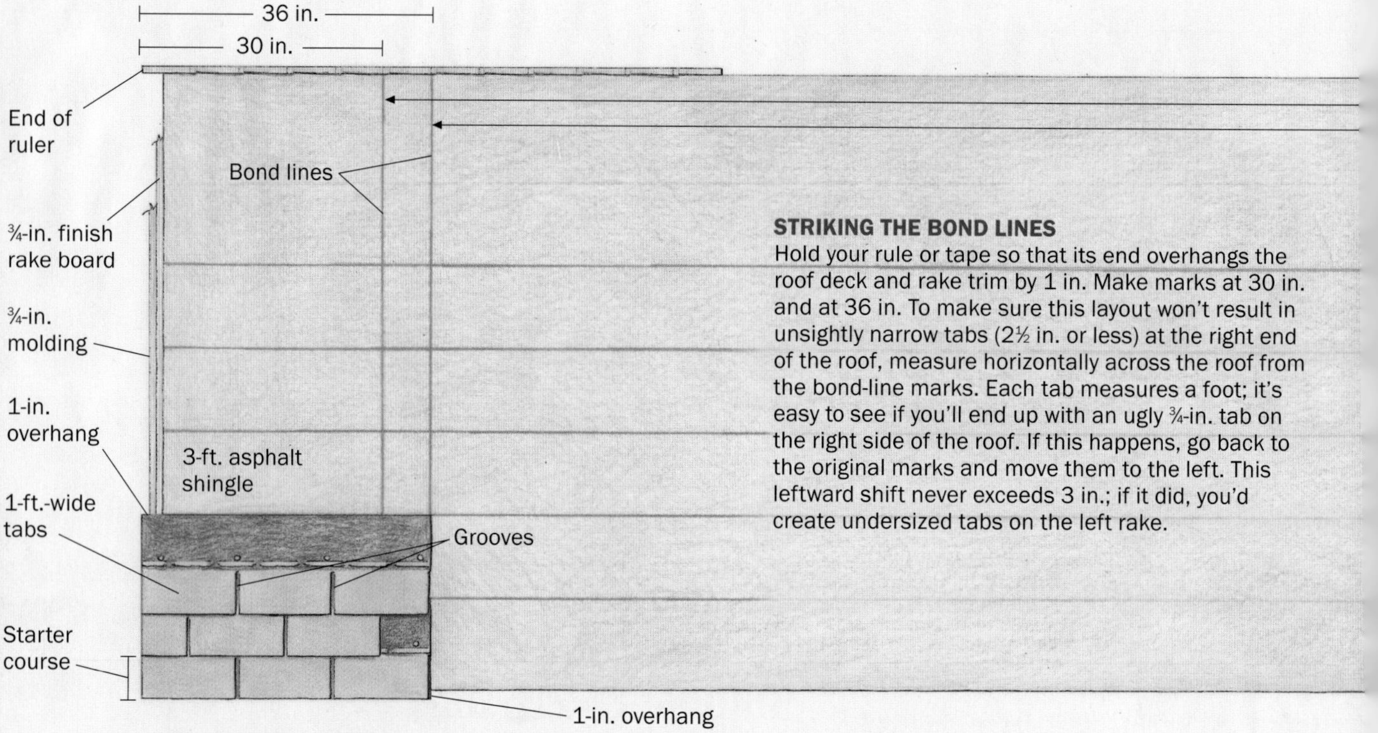

STRIKING THE BOND LINES
Hold your rule or tape so that its end overhangs the roof deck and rake trim by 1 in. Make marks at 30 in. and at 36 in. To make sure this layout won't result in unsightly narrow tabs (2½ in. or less) at the right end of the roof, measure horizontally across the roof from the bond-line marks. Each tab measures a foot; it's easy to see if you'll end up with an ugly ¾-in. tab on the right side of the roof. If this happens, go back to the original marks and move them to the left. This leftward shift never exceeds 3 in.; if it did, you'd create undersized tabs on the left rake.

ESTABLISHING THE OVERHANG
Contrary to manufacturers' recommendations, the author overhangs roof shingles 1 in. on both rakes and eaves. Longer overhangs account for possible discrepancies in the straightness of the trim.

trim pieces and the drip edge will extend the roof deck, and I allow for that extension. For example, if a 1x6 rake board and a piece of molding, totaling 1½ in., will be added to the existing sheathing, I know I should let the shingles overhang the sheathing by 2½ in. to get a final overhang of 1 in. Because shingles are easy to cut, it is better to err on the side of too much as opposed to too little overhang. If need be, I can go back later and trim the overhanging shingles.

Three-tab asphalt shingles are 3 ft. long. They have three 1-ft. tabs with grooves cut between them in the part of the shingle that is exposed to the weather. The grooves both break up the otherwise solid appearance of the shingle (possibly making them look more like wood shingles or slate shingles) and provide a channel for water to run off the roof. It is important that the grooves of every other course line up over one another and that the grooves of the course in between fall in the middle of the tab of the shingle above and below.

Like most right-handed roofers, I usually start shingling on the left side of the roof. This enables me to work from left to right, positioning shingles with my left hand and nailing them off with my right hand.

Striking the Bond Lines

To keep the grooves straight and the shingles properly bonded or centered over the

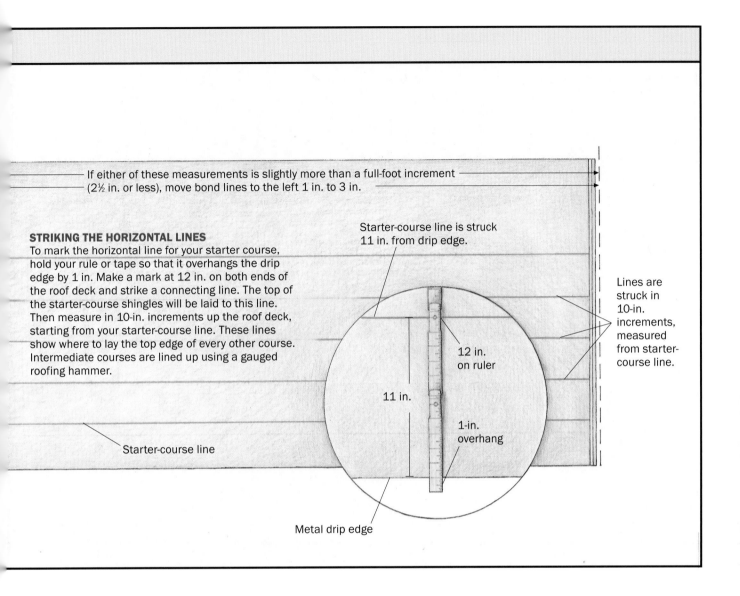

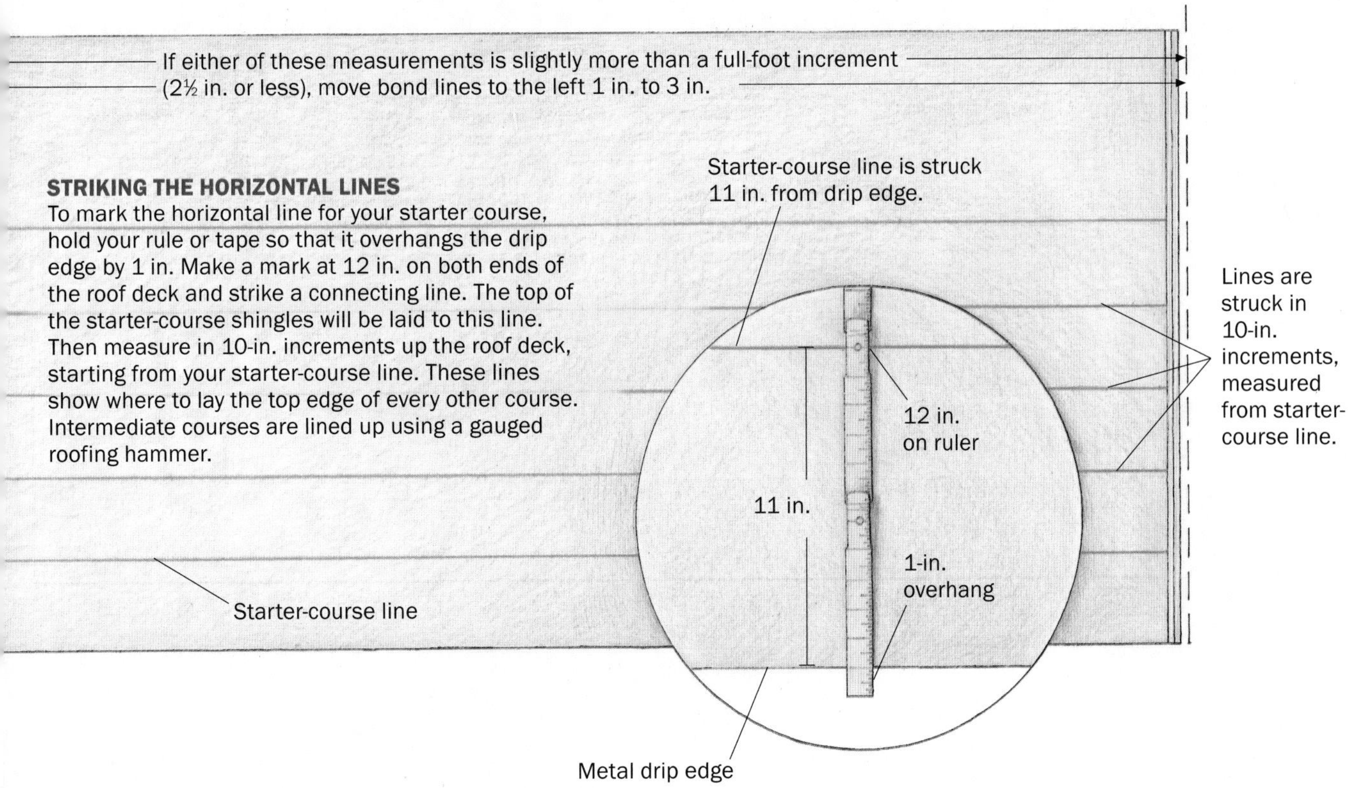

If either of these measurements is slightly more than a full-foot increment (2½ in. or less), move bond lines to the left 1 in. to 3 in.
Starter-course line is struck 11 in. from drip edge.
Lines are struck in 10-in. increments, measured from starter-course line.
12 in. on ruler
11 in.
1-in. overhang
Starter-course line
Metal drip edge

STRIKING THE HORIZONTAL LINES
To mark the horizontal line for your starter course, hold your rule or tape so that it overhangs the drip edge by 1 in. Make a mark at 12 in. on both ends of the roof deck and strike a connecting line. The top of the starter-course shingles will be laid to this line. Then measure in 10-in. increments up the roof deck, starting from your starter-course line. These lines show where to lay the top edge of every other course. Intermediate courses are lined up using a gauged roofing hammer.

tabs just below, the shingles are laid to follow two vertical chalklines, called bond lines, struck near the left rake of the roof (see the drawing above). Bond lines are always struck 6 in. apart—half the width of a tab; this aligns the 1-ft. tabs of alternate vertical shingle rows.

If I need to leave, say, a 2½-in. rake overhang, I extend my ruler exactly 2½ in. past the roof deck and make marks at 30 in. and at 36 in. This is a preliminary measurement. To make sure this layout won't result in unsightly narrow tabs (2½ in. or less) at the opposite edge of the roof, I measure across the roof from these marks. Each foot represents a tab, and it's easy to see if I'll end up with an ugly ¾-in. tab on the right side of the roof. If this is the case, I go back to the original marks and move them to the left. This leftward shift never exceeds 3 in.; if it did I would be creating undersized tabs on the left rake.

When I'm satisfied that I won't end up with little tabs at either end of the roof, I make identical measurements at the top and the bottom of the roof along the left-hand side. Then I strike a vertical, parallel bond line at both the 30-in. and 36-in. marks.

When I'm ready to install the shingles, I begin each horizontal course on a bond line, alternating between the two bond lines. But before I can nail on any shingles, I also have to measure and strike horizontal lines

On a rectangular roof without dormers, valleys, or other obstructions, there are three basic layout steps: establishing the overhang, striking the bond lines, and striking the horizontal lines.

TIP

For those who choose to run shingles vertically, here is one caution: You have to leave the far right-hand nail out of every other course (the one that hits the right-hand bond line).

Striking the Horizontal Lines

Standard shingles are 1 ft. high. To lay out the first course, called the starter course, I need to know the overhang at the roof eaves. If all of the trim and drip edge has been installed, I hold my folding rule so that it extends 1 in. past the drip edge, and I make a mark at 1 ft. I make the same mark at the other end of the roof and then strike a chalkline across the roof deck (see the drawing on pp. 70–71).

Shingle exposure is the height of the shingle that will be exposed to the weather. In most cases, the exposure of three-tab asphalt shingles is 5 in. Shingles are 1 ft. high, so each successive shingle will overlap the one below it by 7 in.

It's not necessary to strike lines every 5 in.; in fact, I always strike lines in increments of 10 in. There's a reason for this, which I'll explain shortly. When I mark my horizontal lines, I place the end of my rule at the starter-course line, and then I make marks every 10 in. If I'm working alone, I often strike lines in increments of 20 in. or 30 in. The most important thing to remember is that all lines are measured off the starter-course line rather than off the drip edge.

Running the Shingles

After striking lines, I start nailing shingles where the bond lines intersect the starter-course line. The starter course is always nailed on the roof upside down. The next row of shingles is nailed right-side up, directly on top of the starter course. The reason for this is to cover the metal drip edge that would otherwise be exposed to the weather by the grooves in the right-side-up second course.

I always begin the upside-down starter course on the left-hand bond line. The next course goes directly on top of the first and begins on the right-hand bond line. Because the lines above the starter course are marked in increments of 10 in., every other shingle hits a horizontal line, and every shingle that does so also hits a right-hand bond line (including the exposed starter). I follow this routine religiously because the consistency is very useful on complex roofs, as we shall see.

Horizontal, Diagonal, or Vertical Shingling?

A neat, professional roof can be installed by running shingles horizontally, diagonally, or straight up the roof. Running each course horizontally across the roof is the simplest method and is usually preferred by amateurs. Running the shingles diagonally across the roof so that they look sort of like a staircase is often recommended by shingle manufacturers because of the possibility that the shingle color might vary from bundle to bundle. The thought is that the variegations will be less noticeable if the different colors are run diagonally rather than straight up or straight across a roof.

Like many roofers, however, I prefer to run vertical rows straight up the roof. I do this for two reasons. First of all, I find it less strenuous because it does not require as much reaching and moving about. Secondly, on hot days I find it to be more comfortable because I'm sitting or kneeling on shingles I've just laid. These are a lot cooler than those that have had a chance to soak up the sun. I've never had a complaint about the blend of colors on any of the roofs I've installed. I have noticed, though, that an off-color bundle looks equally bad whether it runs straight up the roof or diagonally.

For those who choose to run shingles vertically, here is one caution: You have to leave the far right-hand nail out of every other course (the one that hits the right-hand bond line). This allows the shingles in the next row to slip into place. I always use four nails to the shingle in the recommended pattern. To do this, I have to lift the tab of every other shingle in the preceding row.

Using a Gauged Hammer

As mentioned previously, I often strike horizontal lines every 20 in. or 30 in. To keep in-between courses straight, I use an Estwing gauged roofing hammer (Estwing Mfg. Co., 815-397-9558). This hatchetlike hammer has a steel knob bolted through its blade exactly 5 in. from the face of the hammer head (see the photo below). After following the struck horizontal line with one shingle, I line up the next three courses (if I'm using 20-in. increments) with my hammer. The steel knob, or gauge, hooks onto the bottom of the shingle in the previous row, and the bottom of the next shingle sits on the hammer head.

Running straight courses. The author aligns shingles using a gauged roofing hammer and vertical bond lines.

Laying Out Complicated Roofs

So far I've limited this discussion to a straight, rectangular section of roof. Roof planes come in a variety of shapes and sizes, however, and they are apt to be intersected by chimneys, dormers, and adjoining roofs. Shingling around these obstructions complicates the job, but by adhering to a consistent 10-in. layout scheme and using a few simple techniques, it's easy to keep the courses straight and correctly bonded.

To go around a pair of dormers (see the drawing below), I lay out the bond lines and the horizontal courses as previously described. Some of the horizontal lines are

Shingling around Dormers

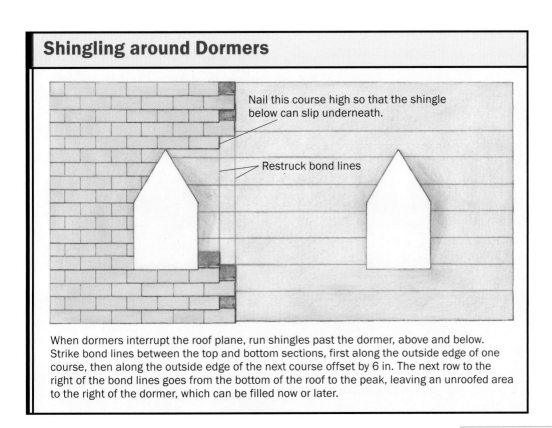

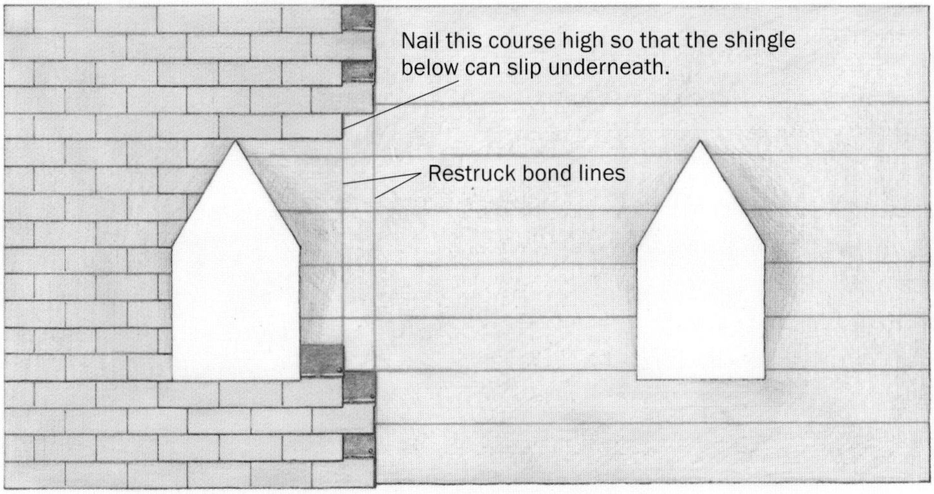

Nail this course high so that the shingle below can slip underneath.
Restruck bond lines

When dormers interrupt the roof plane, run shingles past the dormer, above and below. Strike bond lines between the top and bottom sections, first along the outside edge of one course, then along the outside edge of the next course offset by 6 in. The next row to the right of the bond lines goes from the bottom of the roof to the peak, leaving an unroofed area to the right of the dormer, which can be filled now or later.

Re-establishing the Starter Course

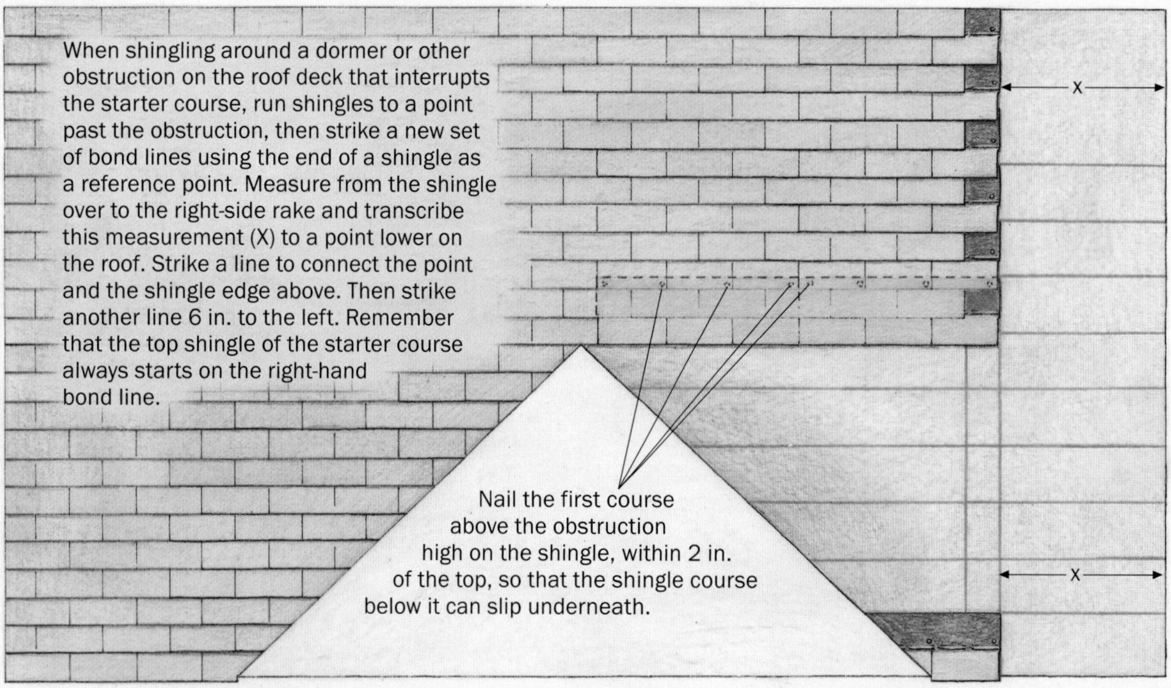
When shingling around a dormer or other obstruction on the roof deck that interrupts the starter course, run shingles to a point past the obstruction, then strike a new set of bond lines using the end of a shingle as a reference point. Measure from the shingle over to the right-side rake and transcribe this measurement (X) to a point lower on the roof. Strike a line to connect the point and the shingle edge above. Then strike another line 6 in. to the left. Remember that the top shingle of the starter course always starts on the right-hand bond line.
X
Nail the first course above the obstruction high on the shingle, within 2 in. of the top, so that the shingle course below it can slip underneath.
X

interrupted by the dormers and have to be measured and marked separately on each side of the dormers. When I start roofing, I run a row of shingles all the way up the left rake and work toward the right until I come to the left side of the first dormer. I continue to shingle the area below the dormer until I'm past the dormer. At this point I move back to the left side of the dormer, cut and fit shingles along the dormer wall, install flashing, and weave the first valley created by the dormer's roof.

There is now a short row of shingles running from the top of the valley to the ridge of the main roof. I carry these courses to the right until they line up with the courses below. To permit the courses that will be installed below these shingles to slide into place, I nail the first course high on the shingle, within 2 in. of the top edge. I strike bond lines between the top and bottom sections, holding the string first along the outside edge of one course, and then along the outside edge of the next course that is offset by 6 in. The next row of shingles goes from the bottom of the roof to the ridge, leaving an unroofed area to the right of the dormer. This area can be filled in now or later, according to the temperament of the roofer. I like to complete this section as I go along.

This process is repeated around the second dormer; I shingle past the dormer at the top and the bottom, strike bond lines through, and fill in.

Reestablishing the Starter Course

When large dormers or intersecting roofs interrupt the bottom section of a roof plane, it is impossible to strike through (see the drawing above). So after running shingles across the top of the roof, nailing the first course high until I've cleared the entire obstruction, I measure the distance from the end of one of the right bond shingles to the right edge of the roof deck. I transcribe this measurement to the bottom of the roof, make another mark 6 in. to the left, and strike bond lines. Then I measure and strike my 10-in. horizontal lines.

I'm now ready to run shingles from the bottom of the roof up the bond lines. But on which bond line do I start? If I pick the wrong one, I'll end up with adjoining courses where all the tabs line up rather than being offset by 6 in.— a roofing abomination. Fortunately, I've struck lines every 10 in., and I've started, as I always do, with the exposed starter shingle on the right bond line. Every shingle that hits a horizontal line also hits a right bond line. I put the inverted shingle of the starter course on the left bond line and cover it with the exposed starter course on the right bond line. As I run up the bond lines, I notice that every shingle that hits a horizontal line also hits a right bond line. I know the bond will work out perfectly.

When There's No Starter Course

Sometimes there's no starter course on the far side of an intersecting roof or dormer (see the drawing below). If so, after I run the top section of shingles over to the rake, I measure and strike bonds in the usual manner. Let's say that I have not struck any horizontal lines in the triangular section created by the intersecting roof. How would I measure down, and what bond line would I start on? To measure down, I extend my folding rule and lay it on the roof deck so that the 12-in. mark is on the bottom of the first shingle in the top section. The shingle is 12 in. high, so this puts the zero point of the rule even with the top of that shingle. Now I mark at every multiple of 10, i.e., 20 in., 30 in., etc., until I get to the bottom of the triangular section of roof. Because

If There's No Starter Course

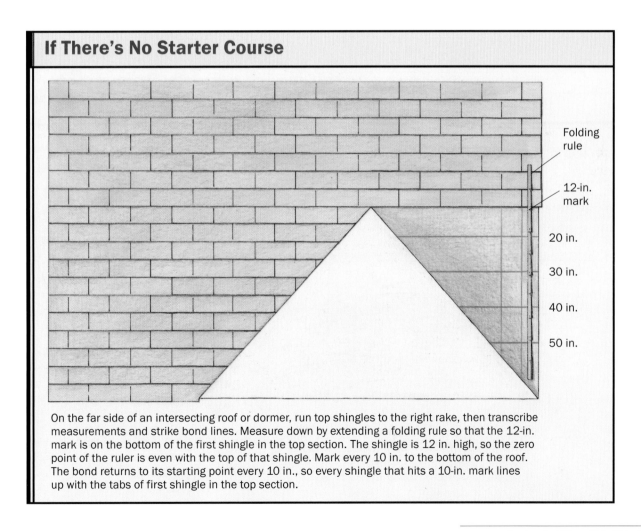

On the far side of an intersecting roof or dormer, run top shingles to the right rake, then transcribe measurements and strike bond lines. Measure down by extending a folding rule so that the 12-in. mark is on the bottom of the first shingle in the top section. The shingle is 12 in. high, so the zero point of the ruler is even with the top of that shingle. Mark every 10 in. to the bottom of the roof. The bond returns to its starting point every 10 in., so every shingle that hits a 10-in. mark lines up with the tabs of first shingle in the top section.

the bond returns to its starting point every 10 in. (or every other course), I know that the grooves of every course that hits a 10-in. multiple will line up with the grooves of the first shingle in the top section. I make sure it does.

Starting a Hip Roof

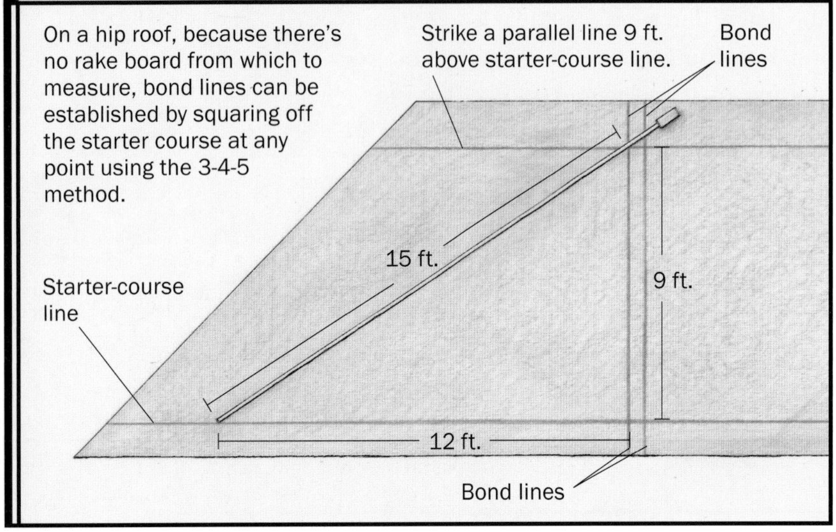

The Slant-Rule Trick

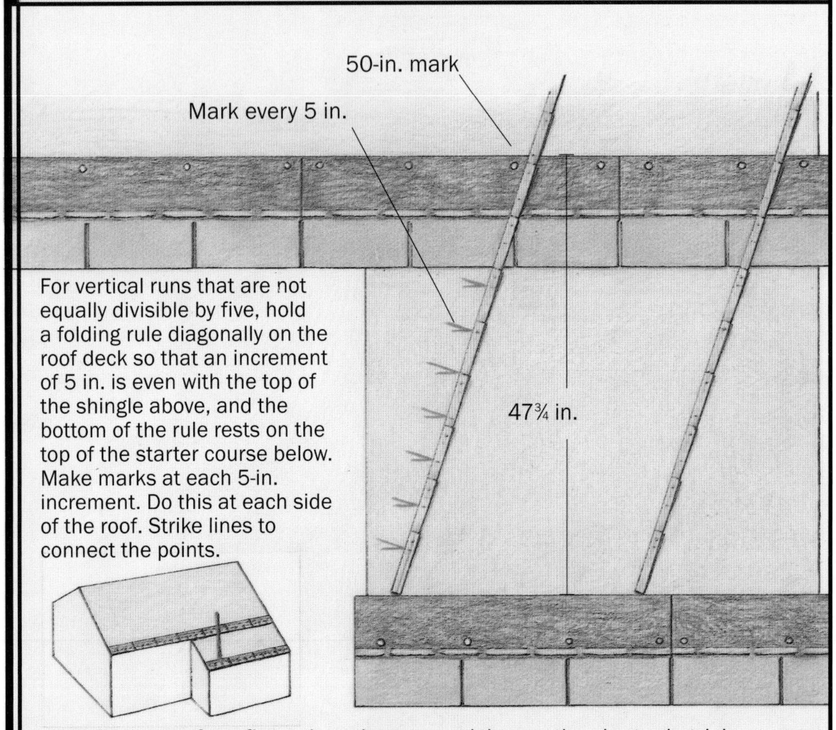

Here's one roof configuration where you might use the slant-rule trick.

Starting a Hip Roof

Sometimes it's not practical to start roofs on the left side. Hip roofs or roofs with obstructions on the left side should be started toward the center of the roof deck. On a hip roof you can't measure from the rake to establish the bond lines. They must be squared up at some point along the starter course using the 3-4-5 method (see the top drawing at left). After striking the starter-course line, I mark a point, measure 12 ft. along the line from that point and make another mark. I strike a parallel line 9 ft. above the starter course, pull a tape measure diagonally from the 12-ft. mark on the starter course until the 15-ft. mark on the tape intersects the upper chalkline, and I make a mark. Stretching a chalkline from my first mark on the starter course to the mark above to the ridge of the roof, I strike my first bond line. I strike a second line 6 in. to the right. Then I can run all of my shingles from left to right, then come back and fill in the hip.

The Slant-Rule Trick

Occasionally it's necessary to fit shingle courses into a space that's not divisible by 5 in. (see the bottom drawing at left). If the run of a section of roof from starter course to intersecting roof is, say, 47¾ in., some roofers might run nine courses at 5 in. and the last course at 2¾ in. A better way to set up the courses is to divide the 47¾ in. into 10 equal courses. Nail on your first two starter courses—inverted and right-side up. Then put your tape on the starter-course line and run the tape diagonally across the roof until you come to a 5-in. increment—in the example above, 50 in. Make a mark at each 5-in. increment. Do this on both sides of the roof and strike lines between each mark.

***John Carroll** is a builder in Durham, North Carolina, and is the author of two Taunton Press books:* Measuring, Marking and Layout *(1998) and* Working Alone *(1999).*

Tearing Off Old Roofing

■ BY JACK LeVERT

Twin brothers Richard and Russell Wright have reroofed hundreds of Boston-area houses since they went into business together 25 years ago. I've been up on the scaffolding with them on quite a few. I usually do the specialty carpentry—gutters, fascia, skylights, sheathing repairs—but when it's time to tear off an old roof, everybody pitches in.

Tearing off a roof (see the photo on p. 78) is a messy, nasty job. You've got to take steps to protect the house, the grounds, and yourself. Here I'll explain how to determine whether a house needs a new roof and what to do if it does.

Inspecting the Roof

When checking out a leaky roof, I first look for structural damage to the roof itself. I do it from inside the attic. I look for signs of continuous moisture, such as water stains, patches of dry rot, or black fungus. And I probe for rot. With an awl, I poke the underside of the sheathing and the top edges of the rafters. Softness indicates water damage. If the interior damage is limited to one spot, this spot may be the source of a leak, and depending on the condition of the shingles, I might simply choose to patch the leak.

If the rafters are sagging, they were probably undersized when the house was built. I jack up sagging rafters and sister on new rafters. This is done before tearing off the roofing; otherwise, jacking up the rafters could spring loose the newly installed roofing or even the sheathing.

Sagging between rafters indicates a problem with the sheathing: It may be rotted or cracked; it may be undersized; it may have been run across too few rafters to provide strength; or it may not have been staggered across the rafters properly. A second layer of sheathing often will correct this problem.

Extensive fascia and soffit damage indicates that either the drip edge is bad, or the rafter tails have rotted. I pull off a section of damaged fascia to check out the rafters. Although I sometimes remove the sheathing to replace rotted rafters, often I can repair rafter tails without removing the roofing. If the rafters are OK, a defective drip edge is probably causing the water damage, and a new drip edge will have to be installed with the new shingles.

If there's any sign of carpenter ants, I tear off some sheathing, determine the extent of the infestation, and get rid of it. The best way to do away with carpenter ants is to remove all wet wood, whether it's infested or not.

Take it all off. When the second layer of asphalt shingles started leaking, it was time to tear off the roofing down to the old board sheathing. After tacking a tarp along the fascia to protect the house and the grounds, the roofers pulled off the shingle caps, then worked their way down with ripping shovels, straight-claw hammers, and pry bars. Triangular brackets are driven between sheathing boards to provide footholds.

What to Look For on the Roof

Like car batteries, shingles are rated to last a certain length of time. Fortunately, shingles are rated in years, not months. You can buy shingles rated to last anywhere from 15 years to 30 years. But you can't determine whether a roof needs to be replaced simply by comparing the age of the shingles to their projected life span. I've heard of shingles deteriorating in as little as five years, and I've seen roofs that have lasted well over 30 years.

You must go on the roof and look for signs of deterioration (see the photo on the facing page)—shingles that have lifted and curled edges, brittle shingles that crumble when lifted. On a cold day, even a new shingle will be brittle and will snap in your hand, but in the old, deteriorated shingle's case, the exposed edges will crack and crumble like stale cake. Here and there a deteriorated roof will look like the worn soles of old shoes, with round holes revealing the shingles underneath. Corners of shingles will be missing, and the heads of roofing nails will show—sure signs of leakage.

In areas that get little or no sun, shingles remain wet for a long time and may be green and slick with moss. Water-damaged shingles will be soft and mushy. If you can leave a thumbprint in a shingle, the shingles are too far gone to keep.

In winter, snow melts slowly in these shaded areas. The constantly trickling water freezes, melts, and refreezes, wearing away

Past their prime. These deteriorated asphalt shingles have missing corners, and roofing nails show. They also snap and crumble easily, and the black showing through indicates that the protective layer of granules has worn away, leaving the asphalt exposed to the sun's ultraviolet rays.

the protective layer of granules on asphalt shingles. The condition of the granules impregnated in the shingles is the real key to determining whether an asphalt roof is shot. Asphalt shingles consist of a blend of steep asphalt (asphalt that won't soften below a temperature of 130°F) held together with strands of fiberglass and covered with a top layer of ceramic-coated granules. Each of these three components has a function. The asphalt protects the roof; the inorganic fiberglass mat strengthens the shingle; and the granules shield the asphalt from the sun. (The granules are crushed rock that resist ultraviolet light, heat up to 1,200°F and acid.) Rain, snow, changes in temperature—all weather—gradually wear away this protective layer of granules. Once the sunscreen of protective granules has worn off, the sun's ultraviolet rays evaporate the natural oils of the asphalt, causing the shingle to degrade.

Roofing Over an Old Roof

In Massachusetts, where we work, the building code allows reroofing over a single layer of shingles but not a third roof over two. This is a good rule to follow, whether you are required to or not. Who knows what the first reroofers covered up. Also, the bumpy surface of two layers of crumbling shingles makes it almost impossible to put down a third roof well. Three layers of shingles might not fit under existing chimney flashing or under the siding of adjacent dormer walls. In some places, curled and buckled shingles will keep nails from penetrating the sheathing; in other places, you'll strike unseen voids and split or tear the new shingles as you nail. Also, it's best to rip off old wood shingles—and even the skip sheathing if it's damaged—and start from a clean deck. If you don't, you could cover up rot when you lay down your new asphalt shingles.

And don't forget that each layer of roofing adds some weight: approximately 2.35 lb. per sq. ft., or 235 lb. per square. (A square is 100 sq. ft.) Each buried layer is an extra burden on the shoulders of an old house. Another consideration is the fire hazard of adding a third layer of petroleum-based material to a wood-frame structure. Most fiberglass shingles are class-A fire resistant; however, no shingle is completely fireproof.

To sum it all up, if I find underlying structural damage, the old roof has to come off. If there are two or more layers of roofing, they must come off in any case. In general, ripping off the old roof makes for not only a more watertight, longer-lasting roof but also a better-looking one. And it makes putting on the new roof easier.

In general, ripping off the old roof makes for not only a more watertight, longer-lasting roof but also a better-looking one. And it makes putting on the new roof easier.

Roofs with dormers, vents, skylights, etc. take longer to tear off and make watertight than, say, a straight gable, so pace yourself accordingly.

Prepare for Tearoff

Before tearing off a roof, we protect the grounds with large plastic tarps. Any windows, doorways, shrubbery, etc. that might get damaged are also covered with tarps. Whenever possible, we nail a tarp to the fascia board or soffit and let it hang to the ground to protect the house from falling debris and dirt (see the photo on p. 78). There are always extra tarps and rolls of roofing felt ready in case the weather unexpectedly changes. (You might pay attention to the weather forecast.) Inside the house, we cover the attic floor with plastic because debris and dirt fall through the cracks in the sheathing—particularly with board sheathing but even with plywood—and mess up the attic.

Picture windows and French doors are sealed with plywood. On the roof, we protect skylights by taping ¼-in. plywood or heavy cardboard over the glass. If we remove a vent or find or cause any other hole in the roof, a piece of plywood is nailed over the hole temporarily so that no one inadvertently steps through it.

Most important, before tearing off the roofing, we decide where to throw it on the ground and how to get rid of the stuff when we're done. On smaller jobs, we use the dump truck. Around here it costs $80* per ton to dump debris at the transfer station.

In many places, you can't bring old shingles to the local dump. Construction debris must be transported to a transfer station that sorts debris into recyclable and nonrecyclable material.

Recycling of roofing material is in its infancy. According to Stuart Laughlin of the Bird Corporation, which manufactures shingles, defective new shingles are now being recycled as a cold-mix asphalt base for roads. The problem with old shingles is that no one has found a way to separate the nails. Bird is experimenting with an electromagnetic process, but for now, unless you can pull every nail, you won't be able to recycle the old roofing.

On bigger jobs, we rent 1 yd. of Dumpster® for every square of roofing. Our Dumpsters come from the private contractor that runs the transfer station. A 30-yd. Dumpster costs $410, which includes delivery to the site and pickup when it's full. If the Dumpster is filled more than once, it costs an extra $65 per ton of material. There isn't much choice when it comes to finding a place for the Dumpster. It's got to be accessible to both the roofers and the carting company, so it usually ends up on the driveway near the house. The driveway's good because a Dumpster will sink into the ground. To minimize driveway damage, we have the Dumpster set on pieces of plywood.

Old shingles will ruin a patch of lawn in a day, so we try to clean the grounds as we go—the safest, most efficient way to work. If you leave the stuff on the ground for a while, at least throw it onto a tarp, cover it each night, and get it into the truck or Dumpster before it becomes a giant, rain-soaked, nail-studded pile.

As long as the roof isn't skip-sheathed, the only difference between tearing off wood shingles and tearing off asphalt shingles is that the transfer station requires wood shingles to be separated from roofing felt for purposes of disposal. Regardless of the material, we tear off old roofs in sections, removing only as much as we can make watertight before the end of the day. Roofs with dormers, vents, skylights, etc. take longer to tear off and make watertight than, say, a straight gable, so pace yourself accordingly.

Getting on the Roof and Staying There

First, we set up staging from the ground to the eaves. This staging may be metal scaffolding, and in places where shrubs close to the house make it hard to set up scaffolding, we set two 2x12 planks on ladder brackets (see the left photo on the facing page). We tie off the tops of the ladders to keep a slid-

Roof bracket. On a steep roof, these brackets provide stable workstations. They are nailed through the roofing into rafters, and a plank is nailed to the brackets through holes at the front of the bracket. Because of the teardrop shape of their nail slots, roof brackets are easily removed by tapping them upward.

Setting up ladder brackets. With extension ladders tied off at the fascia, ladder brackets are hooked onto the ladders about 5 ft. below the eaves. Scaffolding planks—not framing lumber, which is too springy—span the brackets to give you a place to work from when you begin roofing.

Spiking the cleat. A low-tech but effective foothold when working on a particularly steep roof is a 2x4 cleat spiked over the old roofing with 12d duplex nails, which are easy to pull out. These 2x4 cleats are spiked into the roof vertically every few feet, and a row of them continues unbroken across the roof.

ing clump of shingles or a roofer stepping from the ladder to the roof from knocking the ladder sideways.

Once we're on the roof, we nail pairs of roof brackets (adjustable, triangular metal supports for planks) onto the roof vertically every 6 ft. or so (see the top right photo above). The highest brackets are positioned about 6 ft. below the roof's peak, and all the brackets are nailed through the sheathing into the rafters. Planks span each pair of brackets (we use real planks: full 2x12 spruce, not framing lumber, which is too springy), and a roofing nail driven through each bracket into the planks prevents them from twisting or lifting up.

On a particularly steep roof (or when we run out of brackets), we nail 2x4 cleats at the same or smaller intervals up the roof to stand on (see the bottom right photo above). We use 12d duplex scaffold nails, which have a double head, so they're easy to pull out. We always make a continuous row of cleats across the roof with no gaps between them. It is a long first step to the ground.

Safety on the Roof

The first rule of safety on any roof is don't fall off. I wear sneakers. Admittedly, they don't protect against the other hazard—

stepping on nails—but I've found that good cross-trainers provide the best traction on a slippery roof. You may prefer thick-soled boots, but remember that boots are a lot rougher on new asphalt shingles than sneakers are.

One question should govern how you work on a roof: Are you happy there? If you're not happy, you must reevaluate the setup. The most secure staging won't keep a person who is afraid of slipping from doing just that. Trust your fear and make the workstation too safe. Make yourself happy there.

On a typical job with the Wright Brothers' roofing crew, two rippers start work at the peak on the uppermost plank, one person is on the staging below them, and one man patrols the ground. The two rippers tear off the roofing. The person below moves back and forth, taking debris out of the rippers' way, keeping their staging clear and sending the material along to the ground. The ground man, always alert to what's falling from the sky, moves the pile to the Dumpster. The ground man should wear thick-soled boots rather than sneakers. He will be wading in shingles. He should also wear a hard hat. Everyone should wear gloves.

A roofer and a ripper. The main tool for removing shingles from the roof is a shingle ripper, or ripping shovel. Its flat, serrated blade gets under shingles and around roofing nails. The handle is angled steeply so that when you push down on it, you loosen both shingles and nails.

Tooling Up for the Job

When tearing off shingles, we carry 16-oz. straight-claw hammers. The easier the material comes off, the less we use the hammers. But they will be ready to use on nails that remain in the sheathing and for patches of cemented shingles that can't otherwise be dislodged. The straight claw is the roofer's hammer. The story you hear now and again of a roofer who began sliding off the roof and stopped his fall by swinging the claw end of the hammer into the sheathing and hanging on is true.

The ripping shovel, or shingle ripper, is the main tool (see the photo at left) for removing shingles. It's similar to a long-handled spade, except the blade is completely flat with large serrations across the tip, and the handle is steeply angled to give extra leverage for prying up the shingles. You don't remove shingles the way you would shovel dirt. You drive the ripping shovel under as many courses of shingles as possible, and by pushing down on the handle, you spring the shingles loose from the sheathing. The roofing nails hook in the blade's serrations just as they would in the claw of a hammer and are pried up along with the shingles.

Most roofing nails come up with the shingles. The recalcitrant nails are removed later in the final cleanup before papering in the roof (the process of rolling out and fastening felt paper to the sheathing).

Almost as good as the ripping shovel is a regular garden pitchfork. On an old roof with board sheathing, the prongs tend to stab into the seams between boards, but if you keep the pitchfork about parallel to the roof as you drive the fork under the shingles, it will work fine.

That's about it for tools. Pneumatic and gas-powered ripping tools are available, but I've never used them. A ripping bar, a heavy scraping tool for removing tar-and-gravel roofs, may be used to get up cement around chimneys, but usually cement clings to shingles, and leftover clumps of it can be knocked off with a hammer. Later, I'll need a flat pry bar, but for now it's just pitchforks and ripping shovels.

Finesse the flashing. Be careful when tearing off shingles around chimneys, vents, or dormers because you can damage the flashing or the structure. So use a shingle ripper and get close, but pry up the flashing and the shingles around it with a flat pry bar. If the flashing is in decent shape, you might want to save it.

Ripping and Tearing

There are no real tricks to ripping. Once the first patch of bare sheathing has been exposed, we work out in all directions from it. Sometimes a section will come up easier by prying from below; other times we can pry a section off by standing above it. I get the shovel under as many layers as possible and try to spring the nails loose. Just digging and ripping won't do the job; small clumps of material will come up, but most of the shingles will stay nailed down. I save a lot of work by prying as many nails loose as possible and releasing as many layers of shingles at the same time as possible.

Then again, we're careful not to pull up too much at once. It can happen that, once started, the whole roof begins to come up in a vast sheet. I break it into sections about 3-ft. square so that when I throw the shingles down, I won't heave myself off the roof with them.

Loose shingles are slippery and dangerous. I remove as much as I can safely reach from along the roof brackets or the cleats and clean up or have my cleanup person finish that section while I move to the next area. Even if I'm momentarily knee-deep in shingles, I always stand on a clean plank or a solid part of the roof—never on the loose shingles. I keep a push broom handy to sweep the roof and the staging. The granules from the shingles are particularly treacherous.

Working Around Flashing

It's best to leave chimneys, vents, skylights, roof-to-wall intersections, and valleys until last. Here is where most damage to the underlying structure probably is, and it's where most damage can occur if you're not careful about tearing off the roofing here.

Using a pry bar and a hammer, I pry up these last remaining shingles (see the photo above). Around old vent pipes, there's an iron-ring weather seal. I break it off with a few hammer blows and slip an aluminum-flanged flexible rubber boot over the vent pipe. The flange is nailed along the top, and a course of shingles is tucked under the bottom flange. Around chimneys, I very carefully chisel away any old roofing cement with the pry bar and the hammer and pry up the chimney flashing and counterflashing—without removing or damaging either one. The counterflashing is often made of lead and tears easily. It is set into the mortar between the courses of brick as the chimney is being erected, and its replacement is a job for a mason. I bend the pieces away from the roof without tearing them or disturbing their positions in the chimney.

> *It's best to leave chimneys, vents, skylights, roof-to-wall intersections, and valleys until last.*

I might simply remove step flashing, which is woven into courses of roofing at walls and skylights, but I always think of the consequences: Can I put new step flashing in without removing courses of siding? If the flashing is undamaged—and it doesn't leak, and the courses of the new roof will line up properly with the old flashing—I bend the bottom edges of the step flashing up a bit to clean under it, then I weave the new shingles into it when I reroof.

It's hard to detect small cracks in old valley flashing. Even when it appears intact, it's not worth leaving only to find later that the new roof leaks and to wonder if the flashing should have been replaced when the roof was open. So we remove and replace all valley flashing. Occasionally we leave the original flashing and install a new, wider piece over it. In any case, don't walk on that new flashing because walking on it will cause leaks.

Now we're down to the sheathing. We sweep the roof clean, then go over it carefully, pulling up or pounding in all remaining nails.

We stop tearing off the old stuff when we've cleared a manageable section: a section that can be repaired and made watertight by the end of the day. On a small gable we may be able to tear off and make watertight half of the roof in one day. On a large roof, where staging must be moved often, we reroof the section we've torn off before dismantling staging, brackets, and planks so that we won't have to remount them later. It's possible to cover the whole roof with tarps, removing them each day to work and replacing them each night, but we prefer to complete a section, paper it in, and have it ready to reroof before moving on.

Repairing Damage

A rotted piece of sheathing, even if it's only a small section, should be replaced with a new piece that spans at least three rafters. I find the rafters, remove the nails from the sheathing with a nail puller, set the circular-saw depth to just beyond the thickness of the sheathing, and cut it at the centerline of the two end rafters. Then I renail the old sheathing at the cutlines and put in the replacement piece.

Old board sheathing is commonly ⅞ in. or a full 1 in. thick. To replace a few rotted boards, I use 5⁄4-in. rough spruce ledger board if it's available. Where there are lots of boards to replace, I use ⅝-in. exterior plywood and shim the rafters with ¼-in. or ⅜-in. lattice molding, available in various widths at any lumberyard, to bring the plywood level.

If the original sheathing has shrunk, and there are spaces between boards, or if the sheathing is sagging between rafters, I often put a second layer of sheathing over the first. The second layer is usually ⅜-in. plywood staggered and nailed into the rafters. In particular, I avoid joining the plywood in the same pattern as the old sheathing. Staggering the joints will strengthen an old roof and provide an even, secure nailing surface for the new shingles.

Once in a while, we're asked to put asphalt shingles over an old wood-shingle, skip-sheathed roof. We tear off the wood shingles because it's difficult to apply new roofing over them. Then, if the skip sheathing is in good shape, we install a new plywood or OSB roof deck over it, adding molding at the rakes and the fascia to cover the gap. If the skip sheathing has deteriorated, we tear it off down to the rafters and put down a new roof deck.

If the rafter ends are rotted, we remove enough sheathing to scab the new rafters to the old where the old ones are solid. Opening the roof allows us to scab on new rafter tails without disturbing the interior of the house. We may even be able to leave soffit and trim if they are solid. The general rule for scabbing new overhanging rafters to old ones is to extend the new rafter above its bird's mouth (or bearing point) on the exterior wall plate twice the length of its overhang. For example, a 2-ft. overhang requires a 6-ft. length of rafter, 4 ft. of which runs inside the house. If only the last inch or two of the rafters has rotted, this rule doesn't apply; we just scab new pieces to the solid sections of the overhanging members and then cut off the rotted portions.

Papering In

We paper in and prepare for the new roof, even if it won't be shingled immediately. We paper in from the peak down, making whatever's exposed watertight and then continue tearing off the lower part of the roof. Using 15-lb. roofing felt, we paper right over the peak. On the other side of the peak, we put strapping over the paper and nail it down onto the old shingles. Strapping holds the paper securely and is easily removed when we start tearing off the other side.

Because we start at the peak, we nail the felt at the top and the middle only. When we move down the roof, we simply slide the next underlying course of felt under the upper one and tin and nail through both (see the photo below).

We secure the felt with roofing nails and tins. The tins, called buttons in some areas, are aluminum disks that you nail to the felt. Tins secure the felt to the roof much better than staples or nails alone would. Most lumberyards carry either tins or nails with a big, square washer already attached. Properly tinned and nailed, the felt will remain secure against rain and wind until we shingle. (But don't walk on the felt. You won't stay on the roof long if you do. After all, you're walking on impregnated paper that tears easily.)

Across the bottom 3 ft. of the roof, we put a layer of Grace Ice & Water Shield (W. R. Grace and Co.; www.na.graceconstruction.com)—a polyethylene material coated on one side with mastic roofing cement (see the top photo on p. 86). The shield adheres to the roof sheathing and seals it against water backing up from an ice dam. It does not prevent either an ice dam or the backup, but it does form a watertight barrier. Ice & Water Shield comes in 3-ft. by 75-ft. rolls and costs about $75 per roll.

Roll out the paper. You can save a little extra work and keep from walking on the tar paper by papering from the peak down. Before nailing the bottom edge, tuck the lower course under the upper and fasten the whole shebang with roofing nails and tins, which hold the paper better than staples do.

Thwarting ice dams. To get a watertight seal that protects against leaks caused by ice dams, use Grace Ice & Water Shield. It sticks to the sheathing because it's coated on one side with adhesive. Put a single layer of it at the edge of the roof over the bottom drip edge before papering in. Some roofers will even run Grace Ice & Water Shield in valleys for added leak prevention.

Dragging the magnet. You'll have fun picking up nails and all kinds of junk with this sporty magnetic bar. Pull it around the yard, in gardens, over the driveway, and over the street.

If the roof will have woven valleys with no flashing, we put a length of Grace Ice & Water Shield in the valley before reroofing. Most of the time, we flash valleys first, put a length of Grace Ice & Water Shield along each edge of the flashing, and then install the roofing to extend a bit past the Grace Ice & Water Shield.

Grace Ice & Water Shield sticks to the sheathing, to you, and to itself. It's a little like working with a giant roll of electrical tape. The shield goes on over the bottom drip edge but under the new valley flashing. The material is backed with brown waxed paper to prevent it from sticking to itself on the roll. Two workers can handle the material better than one can. You first peel back about 6 ft. of the paper. Then, with one person handling the roll, the other carefully places the exposed material parallel to the bottom of the drip edge, to which it will immediately stick. Once the shield is secure, the rest of the roll can be peeled from the backing and slowly rolled across the length of the roof.

Final Cleanup

Even though we'll have to clean up all over again when we finish shingling, we clean up thoroughly before reroofing, then clean out the gutters. They will be full, and if they're aluminum, they'll be ready to buckle. We rake out the bushes and the yard and clean the driveway. For now, we leave the plastic in the attic because hammering and walking on the roof will certainly shake down more dirt. Then we break out the rolling magnetic bar (see the photo at left). Mine is simply a 2-ft.-long bar magnet attached to a rope. (A similar tool is available from Haase Industries, Inc.; 800-547-7033.) I drag it along the driveway and around the yard. The bar won't pick up most types of flashing, but it's great at collecting nails.

**Prices noted are from 1994.*

__Jack LeVert__ is a carpenter and author living in Natick, Massachusetts.

Reroofing With Asphalt Shingles

■ BY STEPHEN HAZLETT

Nothing lasts forever. Anyone owning a house long enough or buying an older house from someone else eventually has to consider a new roof. One of the first decisions to be made will be whether to tear off the existing roofing materials or to do a layover, laying new shingles directly over old (see the photo at right). A layover may not be appropriate in every reroof situation. But an experienced, attentive roofer can often save the homeowner thousands of dollars while installing a layover that will last as long as, or longer than, the original roof.

A layover can save time and money. If the framing will support another layer, nest the top edge of new shingles up against the bottom edge of the old to achieve a flat roof.

Not Every Roof Should Take a Layover

For the homeowner, the first drawback to tearing off an old roof is that the risk associated with safely getting the old shingles off the house and hauled away could double the cost of the project. Then there is the risk of rain falling on an exposed roof, which novices often don't consider seriously.

If tearing off the old roof is such a nightmare, why bother? For one thing, your local building code might require it. Roofing materials weigh thousands of pounds. If you're covering 15 squares (one square is 100 sq. ft.), 25-year shingles weigh about 3,500 lb., and sturdier 40-year shingles weigh about 5,000 lb. Double or triple that number if there are already two or three layers.

The rafters or trusses and decking will usually support two layers of asphalt shingles, although three or four are common. Most local building codes allow only two layers: the old existing layer and one new layer. Codes vary, so ask a local official for minimum standards. If I see more than one

Align New Shingles with Old for a Smoother Job

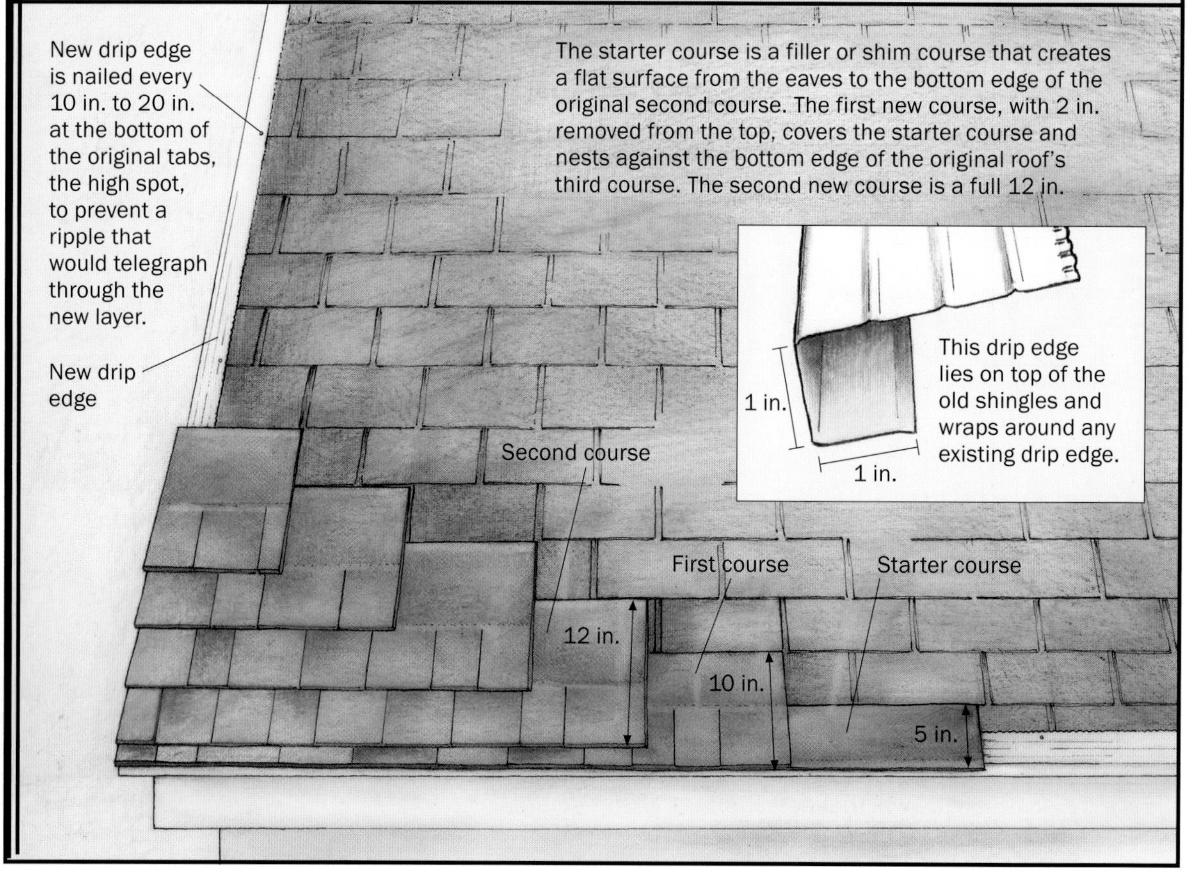

New drip edge is nailed every 10 in. to 20 in. at the bottom of the original tabs, the high spot, to prevent a ripple that would telegraph through the new layer.
New drip edge
The starter course is a filler or shim course that creates a flat surface from the eaves to the bottom edge of the original second course. The first new course, with 2 in. removed from the top, covers the starter course and nests against the bottom edge of the original roof's third course. The second new course is a full 12 in.
1 in.
1 in.
This drip edge lies on top of the old shingles and wraps around any existing drip edge.
Second course
First course
Starter course
12 in.
10 in.
5 in.

existing layer, I recommend a tearoff, even if the municipality involved allows more than two layers, because you can't get three layers to look smooth enough. Also, flashings might be difficult or even impossible to replace, and the old flashings may not last 25 more years even though the new layer of shingles does.

If there's only one layer, I check the existing asphalt shingles for flatness. A lot of curling indicates that even a single layer of shingles will have to be torn off. The uneven surface caused by a curled shingle will telegraph through the new shingles and will not provide good support for the new roofing material. Some cracking of the old shingles is acceptable as long as they are not badly curled.

Tearing off the old roofing also allows me to inspect all roof decking and to replace any questionable wood. I can also install or replace a bituminous membrane to guard against ice dams near the eaves. Because these membranes don't adhere securely to old shingles, they generally don't perform well over old roofing. Also, virtually all the old roof flashings can be replaced during a tearoff, which might not be possible in a layover. If a layover is possible, I carefully evaluate the existing roof vents and flashings.

If the existing roof has one flat layer of roofing and if I am confident that the house can support the weight of a second, I check the home's history of roof leaks and usually evaluate the condition of the decking from the attic. I also walk the entire roof deck, looking for any bouncy areas. Sometimes my crew and I will remove a few shingles in a suspect area to verify the condition of the decking. An extra hour of detective work at this stage of the estimating process can potentially save the homeowner several thousand dollars on an unnecessary tearoff. Finally, faced with a decision of when to reroof, it's always better to do it one year too early than one year too late.

Dimensional Shingles Are a Good Choice for Layovers

One of the last decisions before the roofing project begins is which shingles to use. If a lot of money is being saved by not tearing off the existing roof, it might be wise to invest some of the savings in an upgraded shingle. My labor costs remain the same. When $200* or $300 in materials will make the difference between a 25-year and a 40-year roof, that upgrade gives more value per dollar. Unfortunately, homeowners who know that they will not be living in the same house for 25 years, let alone 40, rarely see it that way.

I have more success selling customers on dimensional, or laminated, shingles such as Elk Prestique® (www.elkcorp.com; 800-650-0355). For aesthetic reasons, dimensional shingles work well in a layover because their texture and lamination help to prevent first-layer blemishes from telegraphing through to the new layer. On a laminated shingle, the nail must go through both sections of the lamination. That's about a 1-in. area through the center. Feel under the shingle as you nail to be sure. I use 1½-in. nails on a layover.

A final word about choosing shingles: The wise homeowner will study warranties. On paper, most warranties are virtually identical, but in practice, they are administered differently. This practice might vary dramatically by region; a respected brand in one locale may be available but poorly serviced elsewhere. A contractor or supplier who regularly handles many different brands may offer valuable advice on which brand has excellent warranty service in your area.

When getting materials to the job site, simple planning can save a lot of effort at a nominal cost. If the site permits it, my roofing supplier is able to deliver shingles to the rooftop. Otherwise, someone will be carrying thousands of pounds, one bundle at a time, up a ladder on his shoulder.

The shingles should be positioned as flat as possible, with the stacks spread along the ridgeline to distribute the weight. Shingles become brittle at lower temperatures and can be damaged if a bundle is bent over the ridgeline. On a steeply pitched roof or in cold weather, I have the shingles delivered to the ground, where they can be stored perfectly flat. If shingles must be delivered to the ground, I get them as close to the house as possible without blocking the customer's driveway or garage and without interfering with ladders or staging.

Prepare the Site

With the materials on site, I can begin work on the roof. First, I sweep off leaves, twigs, or other debris. Ridge vents and hip and ridge caps are removed at this time. If just a few shingles have curled corners, the corners can be clipped off. Occasionally, an entire shingle tab will be removed and a scrap tab nailed in its place to act as a shim. On roofs with plywood or OSB decking, I frequently find at least one 4-ft. vertical ripple in the roofing. Invariably, removing a few shingles in that area reveals one edge of a panel that was not nailed down during the house's construction. I nail down the offending piece, replace the missing tabs, and move on.

Once the roof surface has been cleared and repaired, I install new drip edge. Where conventional drip edge was originally installed, I like to use a drip-edge style we call 10/10 (see the drawing on the facing page). This drip edge has a J-channel profile with a 1-in. by 1-in. bend that wraps around the edge of the existing shingles and drip edge.

If the original roofing lacked drip edge, I cut back any existing shingle overhang so that conventional drip edge can be installed tight against the fascia and rake boards. I nail the drip edge at the high spot at the bottom of a shingle tab, every 10 in. to 20 in. up the rake. The high spot is the bottom of each shingle tab. The low spot is the top of the tab. The difference between the high and low spots is exactly the thickness of each shingle. Nailing at a low spot in the middle or upper part of a tab would cause a ripple in the drip edge that might telegraph through the new layer of shingles.

Cut the Starter and First Courses to Nest New Shingles Against Old

Installing the new layer of shingles begins with the starter course. Remember, the starter course is a filler or shim course; it will be covered and does not count as the first course. Assuming that the existing roof has standard 5-in. exposures, the starter course should also be about 5 in., forming a shim between the eaves and the tab of the second course (see the drawing on p. 88). I cut off the top 5 in. or so of a new shingle, saving the lower 7 in. of this shingle to use later as the highest course on the slope.

The 5-in. strip is laid over the tab of the original roof's first course. Together, that 5-in. strip and the tab of the original roof's second course provide a flat, 10-in.-high area in which to lay the new first course of shingles. Because most shingles are 12 in. high, I cut off the top 2 in. of the new shingles to be used in the first course. (Many dimensional shingles are about 13 in. high, so I cut off the top 3 in. of those shingles to fit them into the 10-in.-high area.)

The second and all following courses are full-height shingles and are butted tight to the bottom edge of the old roof's next course (see the photo on p. 87). This nesting method provides a positive-stop gauge to the work, greatly increases the roofer's speed, and eliminates the snapping of horizontal chalklines.

Failure to use the 5-in. starter course, failure to cut 2 in. off the top of the first course, or failure to use the nesting method will cause the new roofing layer to have a wavy appearance, with the waves running horizontally across the roof. According to the Asphalt Roofing Manufacturers Association℠ (202-207-0917; www.asphaltroofing.org), the nesting method "minimizes any unevenness that might result from the shingles' bridging over the butts of the old shingles. It also ensures that the new horizontal fastening pattern is 2 in. below the old one," which will help to prevent splitting the deck boards.

Replace the Vents as You Go

Most roof flashings and vents will be replaced during a layover. I replace vents and flashings as I get to them.

Slant-back or pod vents are easily replaceable. I remove the nails securing the bottom edge of the vent. Next, I carefully reach inside the vent for a firm grip and quickly hinge it upward. If the vent was properly installed without a lot of tar or caulk gooped over it, this method usually pulls the vent and its nails out from under the shingles quickly and neatly. The hinging action is critical to this trick. I use only metal vents because plastic vents grow brittle over time. Any first-layer shingles torn during this removal are either nailed back down or replaced with scraps of new shingles (see the photo on the facing page).

The replacement vent is installed just as in new construction. Any ridge vents will also eventually be replaced with a style that allows me to nail ridge caps over them, such as Cor-A-Vent® (www.cor-a-vent.com; 800-837-8368).

Warning: Roof vents seem to be great nesting places for wasps. Removing a wasp-filled roof vent 30 ft. in the air on a 12-in-12 roof can really add excitement to your day. Unless you think on your feet a lot faster

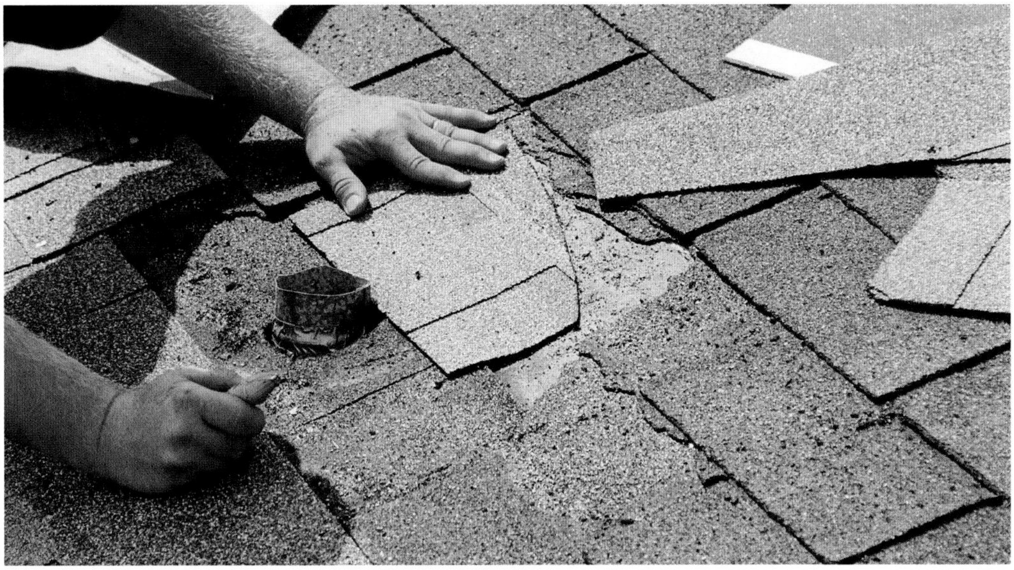

Repair torn shingles around old vents. When a swarm of angry wasps started coming out of this vent, the author didn't have time to remove the old vent gently. He repaired the torn first layer by cutting scraps to act as shims, preventing an uneven second layer.

than I do, have a safe spot picked out ahead of time to toss the wasp-filled vent.

Replace the Flashings

The waste-stack flashing can often be quickly removed using tin snips, a pry bar, and sometimes a roofing hatchet. If a metal waste-stack flashing doesn't look like it will come off easily, I simply cut a few slits in it with my roofing hatchet where the flashing meets the waste stack and flatten it enough that the new flashing will fit over it easily without leaving a visible ripple. This measure is not so much to increase speed as to avoid tearing out old shingles that will then have to be replaced. I usually replace the waste-stack flashing with a metal and rubber unit that gives a tight fit. New waste-stack flashing units come with adjustable rubber gaskets that can be cut to fit different waste-vent stacks (see the sidebar on p. 94). Also, on some older houses with cast-iron stacks, the flared end of the stack may be above the roofline, and an all-rubber waste-stack flashing can stretch over the flare.

Valley flashing is crucial. Luckily, it's also one of the easiest flashings to deal with during a layover because the old valley is simply left in place and the new one installed over it. I prefer to use a metal W valley flashing, and again, as with drip edge, I take care to nail on the high spots.

Some customers prefer the look of a closed-cut valley, in which shingles on one side are cut short of the valley centerline and shingles from the other side cross the valley to underlap those that have been cut (see the drawings on p. 92). Don't be tempted to skimp on flashing here. I have torn off many roofs where the second layer of roofing did not incorporate a new flashing in a closed-cut valley.

Flashing in a closed-cut valley is hidden, so color is not critical. Roll roofing rated at 90 lb. is frequently used for valley flashing, but lately I have been using a modified bitumen product such as Elastoflex® by Polyglass® (800-222-9782; www.polyglass.com). It looks a bit like 90-lb. roll roofing on steroids: thicker and much more durable. That extra durability pays off later when the customer climbs on the roof to retrieve a child's Frisbee and steps right in the center of the valley.

For some reason, roofers often neglect chimney flashing during layovers. Roofers typically will shingle up to the existing

Valley flashing is one of the easiest flashings to deal with during a layover because the old valley is simply left in place and the new one installed over it.

Install New Valley Flashings

Valleys are vulnerable to damage and should have new flashings installed on top of the old valley before the layover.

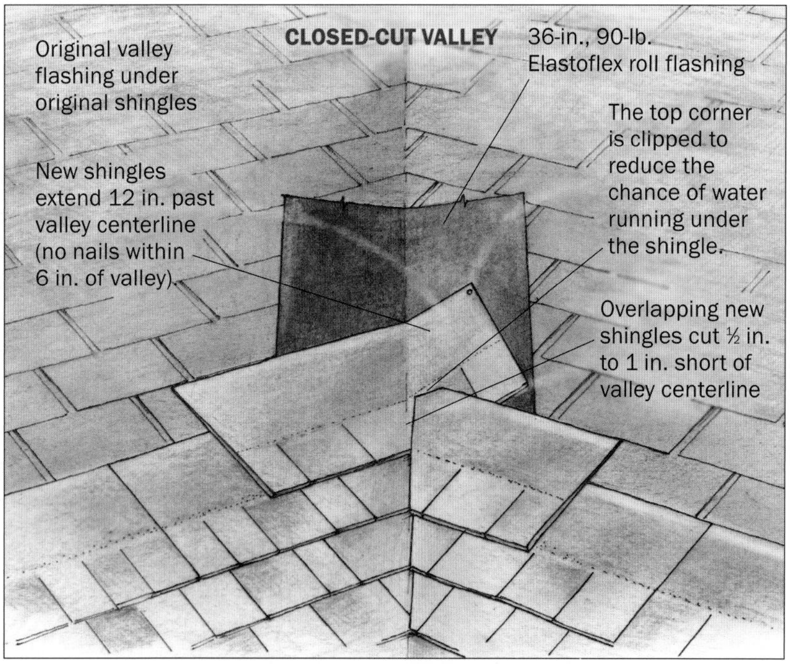

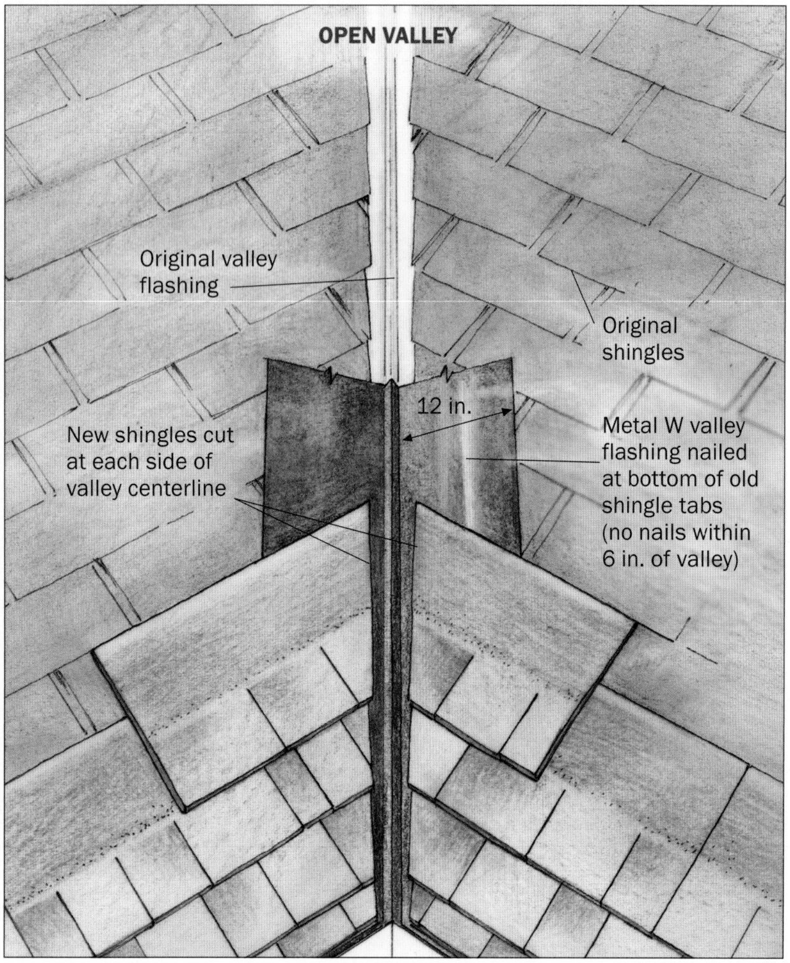

chimney flashing, run a bead of caulk around it, and move on. I like to remove all old counterflashing and caulk. I leave old step flashing in place, but I make sure it is flat on the roof and against the chimney. I install a new apron, step flashing and back-pan flashing. (In this region, we use back pans rather than crickets on the highest face of most chimneys.) Then, after grinding a new kerf (see the left photo on the facing page), I install counterflashing just as on any other chimney-flashing project.

When installing new counterflashing, it would be ideal to reuse the old kerf instead of grinding a new one. But usually, the old kerf is clogged with caulk and concrete nails. Also, a new kerf allows you to raise the new counterflashing higher than the old, thus covering up ancient tar and caulk stains.

Painted aluminum coil is the material most often used for flashing in my area. It comes in 15 to 20 different colors, but brown metal against brown bricks works well on virtually every roof. I do sometimes use black metal on a black-shingled roof.

The most difficult flashing area to replace during a layover is wall flashing (see the drawing on the facing page). Vinyl siding generally accommodates flashing replacement easily because of its flexibility. Aluminum siding is a bit trickier, but still manageable. Wood clapboards in good condition will frequently accommodate slipping in additional flashing, but a few lower nails might need to be removed. In all cases, it's a real help during a layover if the siding installers have left the siding raised a finger's width above the roof. This detail was more common when wood clapboards were the norm because carpenters did not want the siding in constant contact with wet roofing.

Wood shingles, old asbestos, and fiber siding can be brittle and difficult to work new step flashing under, however. I have found old wood-shingle siding by far the most difficult to work with. This material is so delicate that even during tearoffs, roofers often leave the existing sidewall flashing in place rather than risk damage to the siding

New counterflashing. When reflashing a chimney, the author grinds out the mortar for new counterflashing one course above the old. And yes, he should have used a mask or respirator for this work to protect his lungs.

Front-wall flashing. To flash where a lower roof meets the front wall of an upper story, the author first lays the new shingles up to the wall, covering the old flashing. For the last course, he uses the 7-in. shingle left over from cutting the starter course (above). New flashing goes up under the siding and is nailed down.

Don't Rely on Old Step Flashing

Old step flashing can fail just like old shingles. Along sidewalls, the author slips new step flashing under the siding and alternates new flashing with new shingles up the wall. The new step flashing rests on the top portion of the previous course laid, takes one nail in the corner, and is covered by the tab of the next course. New wall flashings are easier to install if the siding contractor has left the siding at least a finger's width above the original shingles.

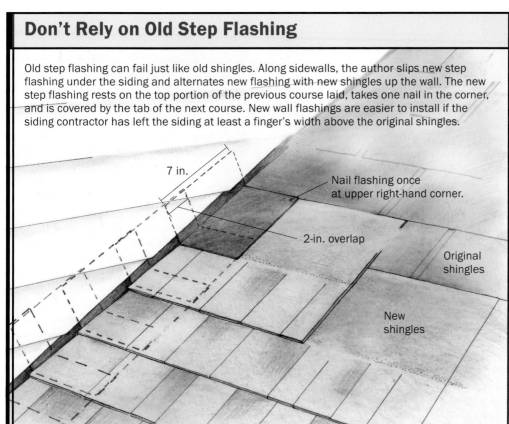

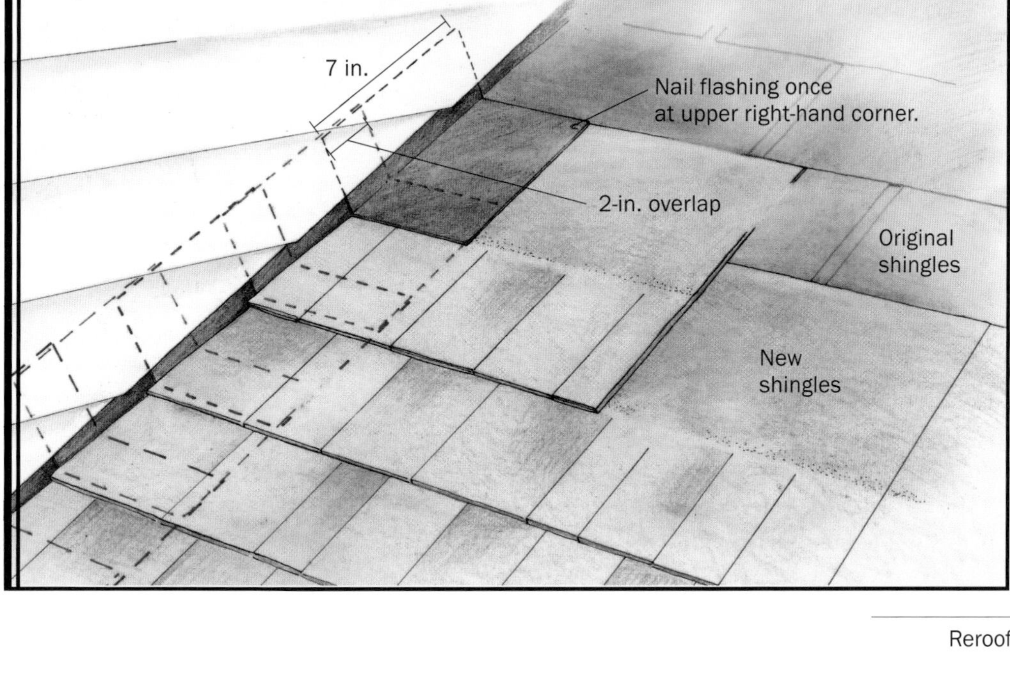

One-Size-Fits-All Waste-Stack Flashing

The author peels away part of the rubber boot to fit the diameter of the vent stack and slides the new flashing over the stack. As new courses overlap the flashing, the author uses a hooked utility knife to cut the shingles so that they lie flat around the boot.

by trying to remove it. Each case is different, and hard-and-fast rules don't apply. This is one area where the roofer's judgment and experience can provide better results than a strict, by-the-book, replace-it-all approach.

One of the great things about a roof layover is that the finish and cleanup work is minimal compared with a roof-tearoff project. Because we don't need a Dumpster on the job and because there are no big trash piles, cleanup is a simple matter of putting wrappers and scraps in trash bags, right on the roof, as we work. Usually, all the job debris fits in one or two standard trash cans, which I empty during our next tearoff project.

**Prices noted are from 2001.*

Stephen Hazlett *owns Hazlett Roofing & Renovation Ltd. in Akron, Ohio.*

Aligning Eaves on Irregularly Pitched Roofs

■ BY SCOTT McBRIDE

Pick up a catalog of stock house plans in any supermarket these days, and you'll see that cut-up roofs—roofs with lots of hips and valleys—are back in fashion. The highly competitive new-home market has compelled builders to spice up their roofs with tasty devices such as Dutch hips and wall dormers. The desired effect is curb appeal, which is the elusive but all-important quality that plays to the homebuyer's romantic notion of what a dream house should look like.

As long as all intersecting roof slopes are inclined at the same pitch, framing a cut-up roof can be fairly straightforward: When the slopes are the same, all of the hips and the valleys run at 45 degrees in plan. Conse-

Spicing up a gable complicates framing and trimming. A pair of steeply pitched gables adds curb appeal to this house with a medium-pitch gable roof. Getting the rafter tails to line up required raising the wall plates on the bays, or kickouts.

While adding steeply pitched features to a medium-pitch main roof might seem like an ideal way to increase curb appeal, it complicates the framing considerably.

quently, all hip-, valley- and jack-rafter cuts can be made on a simple 45-degree bevel (the cheek cuts), and only two plumb-cut angles are required: the common-rafter plumb cut and the hip/valley-rafter plumb cut.

However, combining steep-roofed projections with a medium-pitch main roof is a good way to compromise between cost and curb appeal.

A roof system usually starts with a main gable, and increasing the gable's pitch dramatically increases material and labor costs. Cosmetic roof features such as dormers are much smaller, so increasing their pitch won't have the same impact on cost as will increasing the pitch of the main roof.

Because the usual purpose of cosmetic features is to lend drama to a home's facade, there may be a strong incentive to make secondary roofs steeper than the main roof, especially on the street side.

While adding steeply pitched features to a medium-pitch main roof might seem like an ideal way to increase curb appeal, it complicates the framing considerably. I recently built a house that has such an unequal-pitch condition, and here I want to talk about some of the difficulties I encountered and how I resolved them.

The Particulars of This Job

The house is rectangular in plan, except for two rectangular bays, or kickouts, extending 16 in. beyond the front wall (see the photo on p. 95). The kickouts are topped with 12-in-12 gable roofs. The main roof of the house has a 7-in-12 pitch. Because of the different roof pitches, the valleys don't run at 45 degrees in plan; they're angled toward the lower-pitch main roof. I had to figure out what that angle was and then how to frame the valley.

The eaves were to overhang 12 in. on both the main roof and the kickout gables, and a sloping soffit was to be nailed directly to the rafter tails. If I built the main wall and the kickout walls the same height, the kickout rafter tails would end 5 in. lower than the rafter tails on the main roof (7-in-12 vs. 12-in-12). That would misalign the fascia boards and the sloping soffits.

Complicating matters further, the rafters for the kickout gables were to be 2x6, while the main-roof rafters were to be 2x8. To get a grip on all of these variables, I headed to the drawing board.

Drawing the Cornice Section

I always begin roof framing with a full-scale cornice section drawn on a piece of plywood or drywall (see the drawing on the facing page). In this case I drew one cornice section on top of the other—the 7-in-12 main roof cornice section and the 12-in-12 kickout roof cornice section. The superposed drawing provided me with the length of the rafter tails, the location of the bird's mouths, the width of the fascia, and the depth of the sloping soffits. From the drawing, I also determined how high I'd have to raise the kickout wall so that the kickout fascia would line up with the main fascia.

I began by drawing in the 7-in-12 overhang for the main roof, with the lower edge of the 2x8 rafter starting at the inside corner of the wall plate, the typical location for a bird's mouth. Underneath the 2x8 tail I drew the main-roof soffit. From the point where the face of the soffit meets the back of the fascia, I drew a line at a 12-in-12 pitch to represent the more steeply pitched kickout soffit. Next, I drew a parallel 12-in-12 line from the top end of the 2x8 tail of the main roof; this line represented the top edge of the kickout rafter tail. I now had the top and the bottom edges of my rafter tails aligned at a point 11¼ in. away from the wall. (The ¾-in. thickness of the fascia would increase the overhang to 12 in.)

Next, I drew the 5½-in. width of the 2x6 kickout rafter. The remaining distance between the lower edge of the 2x6 and the

Superposing Main-Roof and Kickout-Roof Cornices

This drawing shows how the wall plate was built up to catch the kickout rafters, which made eaves alignment possible. Starting from the top inside edge of the main wall plate, the author drew the main rafter at a 7-in-12 pitch, then added the 2x6 kickout rafter at 12-in-12 in line with the top end of the main-rafter tail. The bottom of the 2x6 rafters required shims so that the soffits could line up.

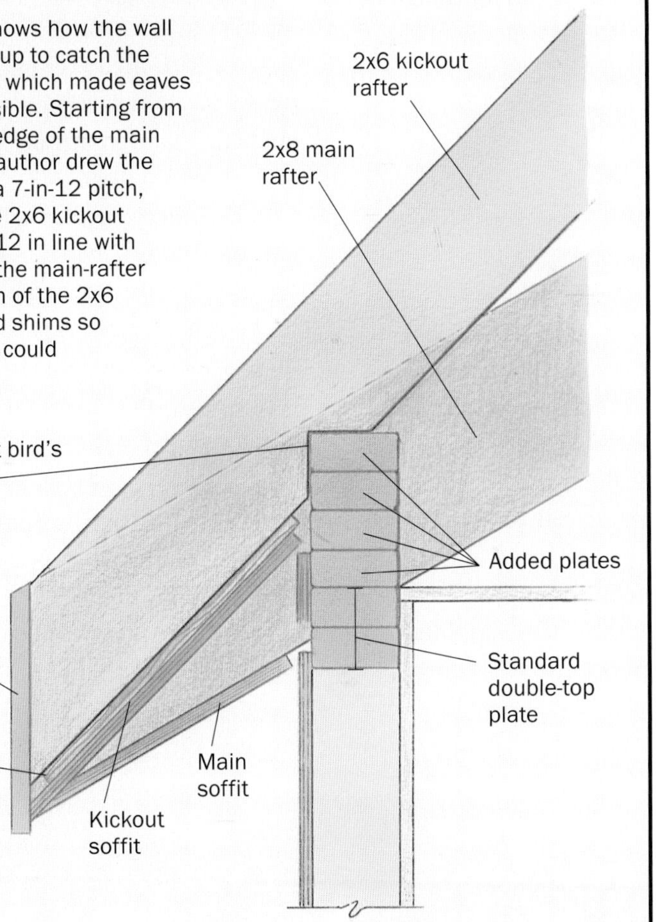

back of the kickout soffit is made up by shimming. The shims cover the underside of the 2x6 kickout tails and extend up along the lower edge of the kickout barge rafters to keep the eave soffit flush with the rake soffit.

At this point I could see roughly how much I needed to raise the kickout wall plate. The exact elevation of the kickout bird's mouth wasn't critical because it's in the attic above the second-story ceiling. I built up layers of 2x4 until the raised plate gave good bearing for the kickout rafters (see the photo at right and the drawing above). With the kickout rafters sitting higher on the wall than the main-roof rafters, the fascia and the sloping soffit could flow in a smooth line from one roof to the other.

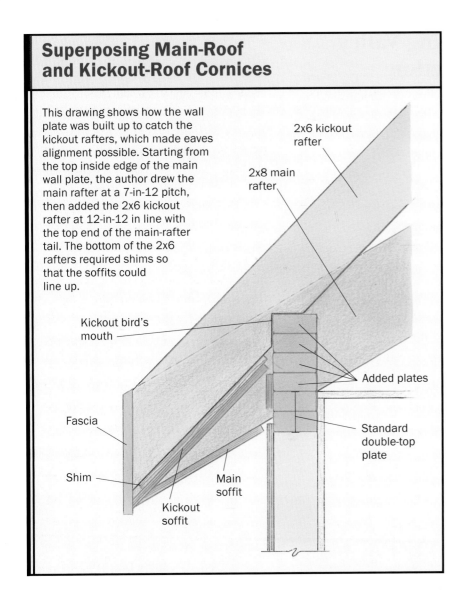

Additional top plates provide bearing for the kickout's 2x6 rafters while picking them up enough to line up the fascia boards.

Simplifying Valley Construction

There are two methods of building valleys. The first, called a framed valley, employs a valley rafter that supports jack rafters coming down from both intersecting roof surfaces. A simpler approach, known as a California roof or a farmer's valley, is to build the main roof all the way across and frame the intersecting roof on top of it. Instead of valley rafters, you nail valley boards flat on the main roof and frame jack rafters for the smaller roof only (see the photo below).

Because a California roof typically sits on the main-roof sheathing, it can't be used if the smaller roof will have a cathedral ceiling. But there were no cathedral ceilings in this house, so I could use the California approach to frame this roof.

Building a California roof simplified the framing significantly because an unequal-pitch valley has several peculiar traits. First, given that the overhang is the same for both roofs, the valley will not cross over the inside corner where the walls intersect, as is usually the case. Rather, it will veer toward the roof with the lower pitch. A valley rafter's location would have to be figured out beforehand from studying a plan view of the roof framing.

Furthermore, an unequal-pitch hip or valley rafter requires two different edge bevels at the point where it hangs on intersecting ridges or headers. One of the bevels will be sharper than 45 degrees, so it cannot be cut with a standard circular saw. With the California roof, I avoided the problem of dissimilar edge bevels and the hassle of locating valley rafters.

I located the off-center valley by snapping lines on the main-roof sheathing. First, I installed the kickout ridge and its common rafters. Then, at the peak of the kickout

One roof framed on another. A California roof features valley boards that lie flat on the main roof sheathing. As opposed to a valley rafter, there's only one set of jack rafters to cut and one cheek-cut setting on the circular saw.

gable, I anchored a chalkline, stretched it diagonally across the kickout common rafters, and marked a point somewhere near the eaves where the chalkline hit the main-roof sheathing. The top of the valley occurs where the kickout ridge dies into the main roof, so, by striking a line across these two points, I located the valley boards. I beveled the edges of the valley boards and nailed them flat on the main-roof sheathing.

Mitering the Soffits

Once I had the kickout jack rafters and sheathing in place, I turned my attention to the cornice. Thanks to the raised plate on the kickout wall, all of the rafter tails lined up, and the fascia flowed smoothly around the corner. But now I had to make the sloping soffits do the same thing.

The main question was, at what angle should the soffit boards be cut to create a clean miter at the inside corner (where the main wall and the kickout wall intersect)? I could cut some scrap pieces with 45-degree face cuts and continue adjusting the angle by trial and error until I had a good fit. Or I could stay on the ground, figure out the angles on paper, and install the soffits on the first try.

I opted for method two, and I accomplished this through graphic development. Graphic development is a way of taking a triangle that occurs in space, such as the gable end of a roof, and pushing it down on a flat surface where it can be measured accurately. All you need is a pencil, some paper, a framing square, and a compass. A stubborn disposition helps, too.

To figure out the miter angles of the soffits, I began by drawing a plan view of the wall lines and the fascia lines (see the drawing on p. 100). I then drew in plan views of the main-roof rafter tail and the kickout rafter tail, each perpendicular to its respective wall. In addition, I drew elevation views of the same rafter tails. I used the numbers 7 and 12 on the framing square to draw the main-rafter elevation view and 12 and 12 for the kickout elevation view. As the plan view and the elevation view of each tail cross the wall lines, they show the vertical rise of the tails: 7 in. for the main roof, 12 in. for the kickout.

Next, I needed to draw the valley. Remember, it doesn't run at 45 degrees. I already had the end of the valley: the point where the fascias intersected. What I needed was another point farther up the valley so that I could draw the valley line. Because I already had drawn the 12-in. kickout elevation view, I decided to find the point where the valley rises 12 in.

First, I extended the kickout-wall line in the direction of the valley. At this line, the kickout roof rises 12 in. above the fascia, so the line represents plan views of both the kickout wall and the kickout roof's 12-in. rise line. Then, I extended the main rafter tail and drew a perpendicular line showing the plan view of the main roof where it rises 12 in. above the fascia. The intersection of both of the 12-in. plan-view rise lines is a point on the valley, and, by connecting this point with the inside corner of the fascias, I drew the valley in plan.

To determine the miter angles (or face cuts) for the soffit material, I used a compass to swing the 12-in.-high elevation view of each common rafter tail directly over the plan view. This point represents the actual length of the rafter where it rises 12 in. (as opposed to the foreshortened length of the rafter when seen in plan).

Then, I drew lines perpendicular to the plan views of the rafters at the points where my compass intersected them. These are labeled elevation lines on the drawing. Next, I intersected the plan line of one soffit with the elevation line of the other soffit. I connected these intersecting points to the inside corner of the fascia, giving me the angle of each soffit's face cut.

Try to imagine the inside soffit edges rising up while the outside soffit edges remain "hinged" along the fascia line. When the

Combining Plan Views and Elevations

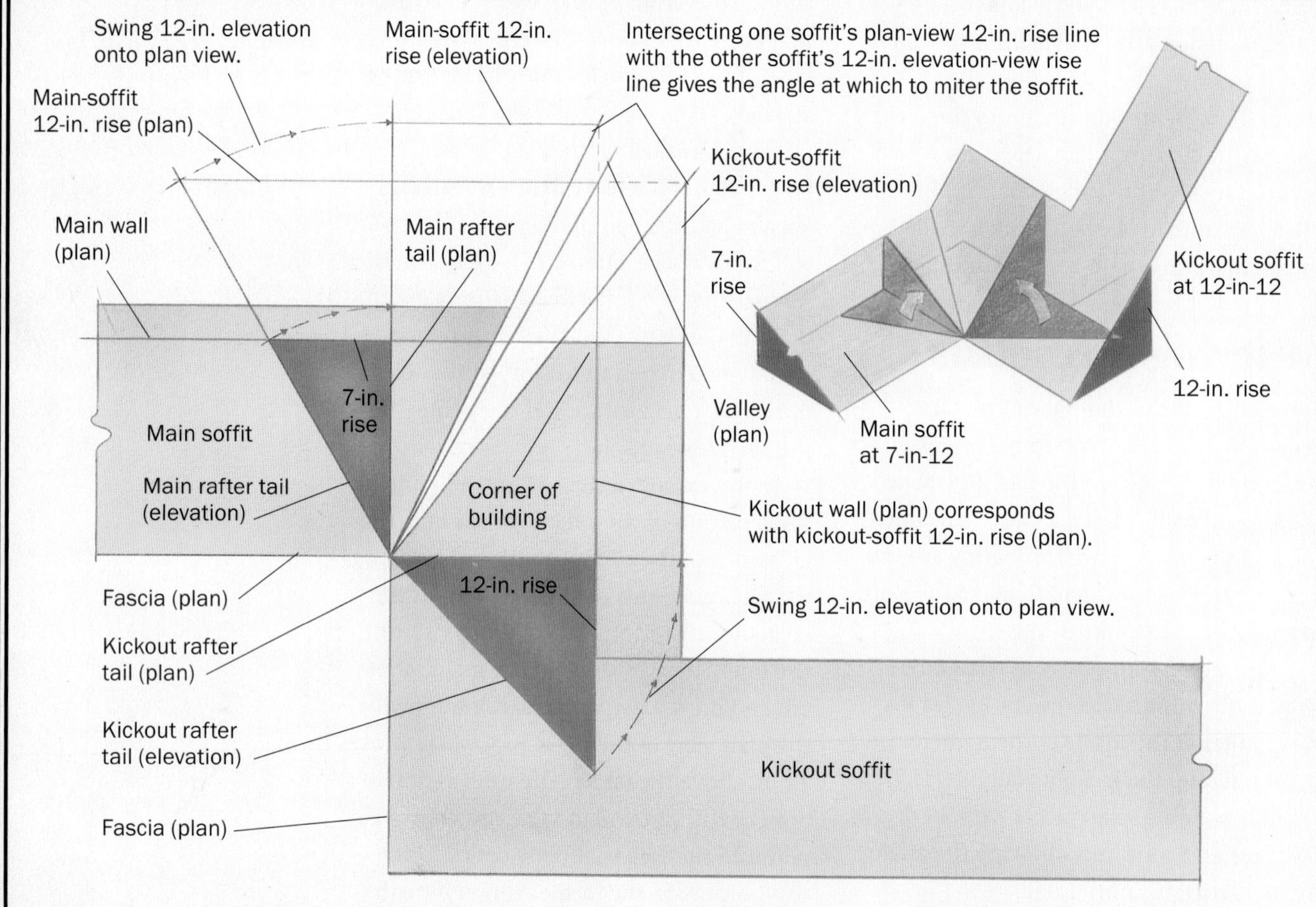

The challenge was to miter the soffits so that when they're nailed to the rafters, the soffits intersect along the valley line. The solution was found through this graphic development (lower drawing), which shows the soffits lying flat on the paper. The first step was to draw the walls and the fascia. Next came plan views and elevations of the rafter tails. Then, with a compass, these elevations were swung back onto their plan views, which gave the widths of the soffits. Finally, the angle of the soffit miters came from intersecting the plan view of one soffit at a given rise—in this case, 12 in.—with the 12-in. elevation line of the other soffit. When the soffits are pitched at 12-in-12 and 7-in-12, they come together directly above the valley line in plan (upper drawing).

inside edges have risen 12 in., the soffits touch along their face cuts, forming a valley, or as viewed from below, an upside-down hip (see the photo on the facing page). The meeting takes place directly over the plan view of the valley line.

The face cut for the kickout soffit was the same as the face cut for the kickout roof sheathing, which happened to be 1x6 but could just as easily have been 4x8 sheets. The sheathing sits on top of the rafters, and the soffit hangs below. Otherwise, they're the same.

Because of its steeper pitch, the kickout soffit dies square into the house for a short distance before mitering with the main-roof soffit. I could tell from the graphic development where to cut the kickout soffit along the main-wall line. The pieces fit on the first try.

Joining Mitered Soffits

The edge bevel for the soffits wasn't critical because the back side doesn't show. I just cut them at 45 degrees, which undercut the

Sloping soffits make for tricky miters. One problem with this roof was getting the sloping soffits of two different roof pitches to flow smoothly around the inside corner. The miter mirrors the offset valley, so the kickout soffit dies square into the main wall. A beveled 2x4 provides backing along the miter joint.

pieces more than necessary and assured a tight miter. Fastening the intersecting soffits presented a problem, however, because I didn't have a valley tail to nail the ends of the soffit plywood to. (That would have been the only good reason for using a framed valley here instead of a California valley.)

To solve the problem, I connected the main-roof soffit and the kickout soffit along their intersection with a beveled 2x4 backerboard. I beveled the 2x4 to match the angle of the valley trough. The 2x4 backerboard didn't have to fit tightly between the fascia boards and the house because the ⅜-in. soffit material was pretty stiff. Instead, I just cut the ends of the backerboard for a loose fit and pulled it against the soffits with galvanized screws.

***Scott McBride** is a contributing editor of* Fine Homebuilding. *His book* Build Like a Pro: Windows and Doors *is available from The Taunton Press. McBride has been a building contractor since 1974.*

Installing a Rubber Roof

■ BY RICK ARNOLD AND MIKE GUERTIN

Not long ago, we'd cringe whenever we'd get a job that had a flat roof or a roof with a really shallow pitch. Add a couple of extra details such as a 6-ft. French door opening onto a wooden deck over the roof with kneewalls on two sides, and we'd hear voices screaming in our sleep: "I'm going to leak, I'm going to leak."

The only way we could guarantee a watertight job was to have a copper roof pan fabricated to cover the flat part of the roof. But copperwork isn't cheap, and we'd still have to do some fancy flashing. In most cases, we were forced to fall back on that old inferior standby, roll roofing. Because of roll roofing's poor track record, we always left the homeowners a bucket of tar at the end of the job. Eventually, they'd need it.

Then, about eight years ago, we installed our very first EPDM (ethylene propylene diene monomer), or single-ply rubber, membrane on a large flat roof, and since then the voices of doom have all but disappeared. Rubber roofing, used commercially for many years, is now finding its way onto more and more residential projects. Properly installed, a rubber roof can solve even the most difficult flashing details. And unlike with roll roofing, we've never been called back for a rubber-roof job that leaked, even on oceanfront projects that experience gale-force winds on a regular basis.

EPDM Membrane Is Sold by the Yard

There are many different systems for installing EPDM membranes, including loose-laid and ballasted (where the membrane is put down without being attached directly to the roof); mechanically fastened, hot-applied fully adhered (also known as the torch-down system); and fully adhered. Each system has different performance characteristics that make it suitable for specific applications.

We primarily use the fully adhered system for installing rubber roofs (see the drawing on the facing page). This system is the most cost effective and easiest for us to install by ourselves on the small-area roofs that we're asked to do most often. Fully adhered installation systems don't require special tools, and decks can be installed on top of them.

The first task is determining how much and what kind of rubber-roof membrane a job requires. The membrane we buy from a local roofing-supply company typically comes in 10-ft. by 50-ft. or 10-ft. by 100-ft. rolls, although some manufacturers offer

Putting Down a Fully Adhered Membrane

A sheet of cured membrane is glued to the roof deck and wall with contact cement. A length of aluminum trim stock covered with membrane secures the edge of the EPDM where it overlaps the edge of the roof shingles. The decking is attached to sleepers atop EPDM chafing strips.

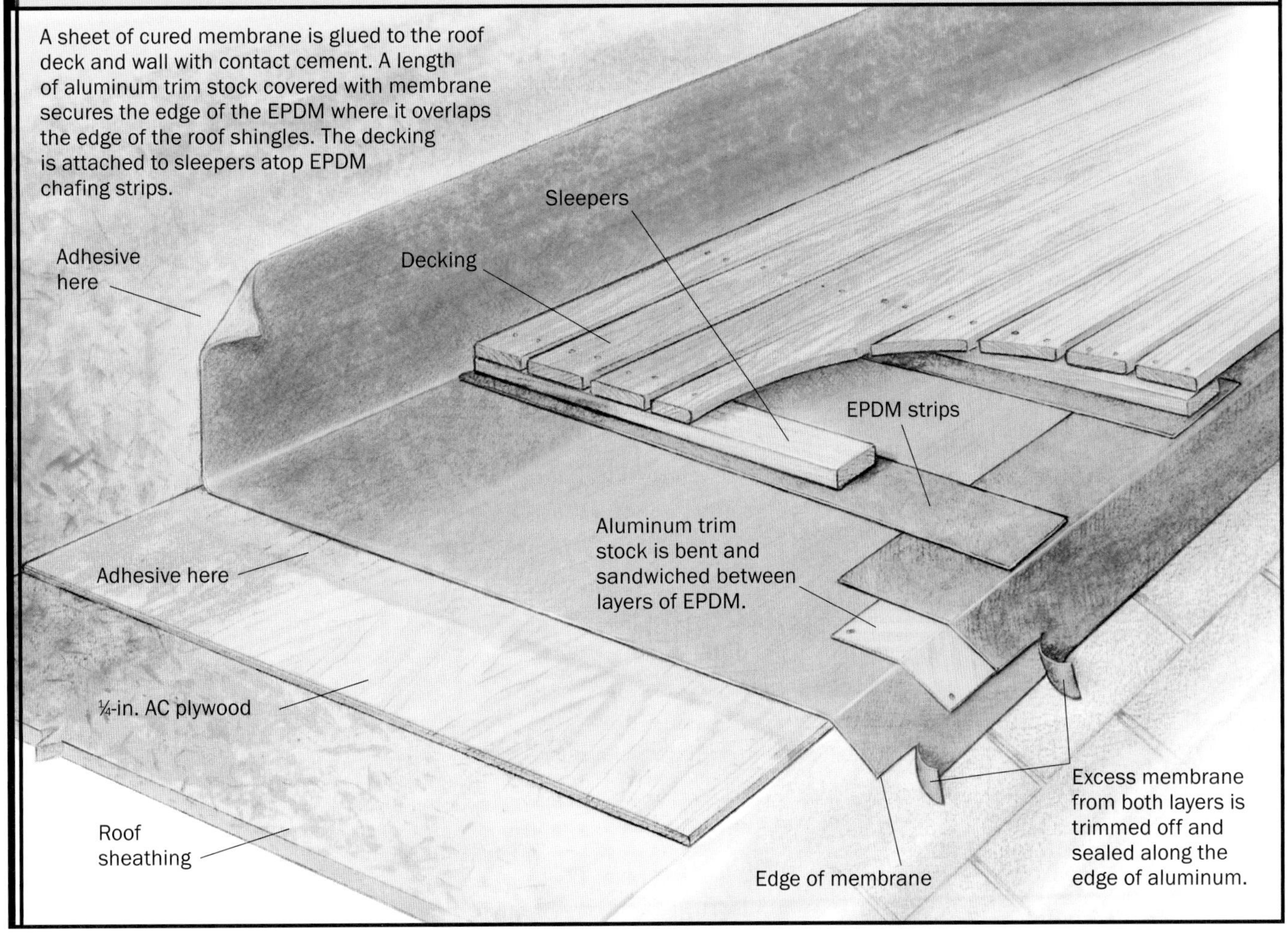

various lengths and widths. For a small job such as the one in this article, we pay a few cents more per square foot to get the membrane cut to the length we need. The extra cost is offset by not having a lot of leftover material.

The membrane is also available 0.045 in. thick or 0.060 in. thick. However, for a fully adhered system, manufacturers recommend the 0.060-in. membrane because a tiny bit of the membrane is supposedly dissolved when the adhesive is applied. Also, for a job that's going to see traffic or have anything put on top of it, the extra thickness is insurance against roof failure.

EPDM also comes either cured or uncured. The membrane that covers most of the roof is made of cured rubber, which is stiffer and stretches less than uncured rubber. Uncured membrane is sold in narrower strips with a peel-off backing. Uncured membrane is flexible so that it can be stretched around corners and over seams to flash in the membrane and make it continuous and leakproof. But uncured membrane will also deteriorate over time when left exposed, so we use as little as possible.

The Work Area as Well as the Substrate Must Be Kept Clean

Before measuring and cutting the membrane, we prepare all areas that will receive the rubber membrane, including the flat

The first task is determining how much and what kind of rubber-roof membrane a job requires.

roof, intersecting walls, door thresholds, adjoining roof sections, and, in this case, kneewalls. All sheathing must be securely fastened, and all fasteners have to be sunk flush or below the level of the sheathing. We also check for sharp edges that might puncture the membrane.

The list of substrates approved by EPDM manufacturers includes plywood, OSB, wood planking, and even lightweight concrete. Most EPDM membranes can also be installed over polyisocyanurate insulation board and high-density fiberboard panels placed over just about any other type of solid substrate. Insulation panels are also available in different thicknesses for different R-values, and they are often sold by the same manufacturers that make EPDM. On this job, insulation was not required, but the OSB roof deck had been exposed to the elements for a while, and the many loose surface flakes would have compromised a connection with the EPDM. To give us a fresh surface for the best bond with the EPDM, we screwed down a layer of ¼-in. smooth-sanded plywood on top of the OSB.

As a final preparation, we give the substrate a good sweeping or vacuuming. All areas that will come in contact with the membrane must be free of dirt, dust, and debris. The membrane itself has to be kept clean. Any small particles that can get trapped under the membrane may cause it to fail prematurely.

We also prepare a large, open work area for measuring and rough-cutting the membrane. For this project, our work area was the plywood subfloor of a large room just inside the future rooftop deck. After taking rough measurements of the deck and the adjoining kneewalls, we carefully rolled out the membrane and cut it to the approximate size. This roof deck was around 13 ft. wide, including the foot or so that we let extend over the roof shingles and the extra couple of feet that run up the walls. Because the membrane came only 10 ft. wide, a second piece would be needed to finish covering the roof.

Unrolling the membrane for a dry fit. After the substrate has been prepared properly, the sheet of EPDM is laid out in the area to be covered for a dry fit.

Corner Cuts Don't Have to Be Precise

We sweep the roof area one last time and then lay the large cured-rubber membrane down for a dry-fit trim. Scissors or a razor knife works great for this part. We're not too fussy when it comes to trimming the extra membrane at the various corners. Those areas will be covered with additional layers of flexible uncured rubber. And if a wrong cut or an accidental tear is made, the membrane can be added to and repaired easily. Rubber roofing is forgiving.

We glue the membrane to the deck floor before gluing it to the kneewalls. So once the trimming is complete, we prepare for the glue-down by keeping the membrane in its dry-fit position and folding the material back in 2-ft. to 3-ft. increments from one edge. We try to work with only a small section of membrane at one time so that it's easy to reach both the substrate and the membrane when spreading the adhesive (see the top photo on the facing page). For this job, we first folded back the material that was going to run up one of the kneewalls. Then we folded back an additional 3 ft. or 4 ft., exposing the first section of roof deck along that kneewall.

Membrane is not glued down all at once. The adhesive for attaching EPDM is a type of contact cement. Folding back the membrane into smaller sections makes it easier to apply the glue and to stick down the membrane.

We use two adhesives for gluing the membrane, both of which are contact cements. One adhesive is for gluing EPDM to the substrate. The other type that we use later can also be used to glue EPDM to itself. To prevent problems with incompatible materials, we always use adhesives specified by the EPDM manufacturer.

We begin our glue-down by applying adhesive to the exposed portion of the deck with a paint roller. We brush the glue onto the corners as well as under the fold of the membrane to make sure we get proper coverage. Next, adhesive is applied to the membrane that was exposed when it was rolled back. We always double-check the coating on the edge of the fold to make sure there are no voids.

The two surfaces are ready to be mated when the adhesive has dried to the touch. The adhesive dries more slowly the heavier it's applied, and bonding the membrane over wet adhesive causes the membrane to bubble. So we always make sure the thickest areas are dry before proceeding. At that point, the membrane is rolled back down slowly (see the photo at right).

Where the rubber meets the roof. After the adhesive has dried, the crew unrolls the membrane slowly, making sure no wrinkles or air bubbles remain. Once it's unrolled, every inch is pressed down to ensure good adhesion.

A roller presses down the corner. After tacking the flaps of kneewall membrane up out of the way, a roller is used to press the EPDM all the way into the corner between the deck and the kneewall.

Working two areas at the same time. Because chilly weather slows drying time, adhesive can be spread on the first kneewall while the glue on the last deck area is drying in the foreground.

The surfaces bond immediately and permanently as soon as they touch, so we try to smooth out wrinkles or air pockets as we go. Most of the smoothing is done with the palm of the hand, but we also use a small roller to press down stubborn spots. When the membrane is stuck down as far as the kneewall, we tack the kneewall flaps loosely in place and press the membrane into the corner with a roller (see the photo at left).

Next we go to the opposite wall and fold up the rest of the membrane until the next section of the roof deck is exposed. Because the membrane is the farther away of the two surfaces, we kneel on the substrate and spread the glue on the membrane first and then spread the glue on the next section of roof deck. After the second section is pressed down, the process is repeated until we reach the other edge of the roof deck. We let the excess membrane run out onto the roof shingles to be dealt with later.

Membrane Continues Up the Kneewalls

On the job featured here, we put the membrane down on a cold day in November, and cold temperatures slow the adhesive's curing time. So while we were waiting for the adhesive to dry on the last section of roof deck, we began prepping the kneewall on the opposite side.

With kneewall flaps tacked in place, we trimmed off excess membrane in a level line around 18 in. up the wall. Then we drew a pencil line along the top edge of the membrane so that we knew how far to spread the glue. The kneewalls on this job were sheathed in OSB, but we weren't concerned about minor flaking of the board. The membrane was to be covered with sidewall shingles that would help to hold it in place.

After we spread the glue on the kneewall (see the photo at left) and the EPDM flaps, we had to be very careful with the flaps while they were drying. The kneewalls had an angled jog to them so that when the flaps were folded back down, they overlapped and had to be kept apart while the glue was drying. To separate the two flaps, we draped one of the sections over a short piece of 2x placed on edge.

Working from the bottom up. When pressing the kneewall membrane into place, the top portion is held away while crew members start from the bottom corner and work their way up.

While we were waiting for the glue on the first kneewall to dry, we stuck down the last part of the roof deck and spread glue on the other kneewall. We began sticking down the kneewalls by making sure the membrane was pushed all the way into the corner between the roof deck and the kneewall (see the photo above). Some projects call for a wooden cove to be installed in the corner, which creates a softer transition between floor and walls. Then we worked the membrane up the wall slowly one flap at a time. When the membrane on the walls was completely stuck down, we pressed the membrane into the corners with the roller.

Deck Membrane Is Finished With a Splice

As we mentioned earlier, the membrane stock was not wide enough to cover the roof with its overlaps onto the roof shingles and the wall of the house. So our next step was splicing on a second piece of EPDM to complete the roof.

We like to have at least a 6-in. membrane overlap, so we measured out that distance from the edge of our first piece and snapped a line. Then we measured from that line to the corner and up the wall about the same distance as our kneewall coverage. We swept the work area again and cut out the membrane we needed.

As with the first piece of EPDM, we dry-fit the splice, aligning it with the line we'd snapped for the overlap. Because the first piece came within a couple of inches of the house wall, we cut our splice piece to fit between the kneewalls, eliminating a difficult corner detail. The small strip of kneewall left exposed would be covered with the uncured EPDM flashing. We marked the door opening on the EPDM so that we spread only enough glue to wrap the step and the threshold.

For the splice, we applied the membrane in reverse, gluing the membrane to the house wall first (see the top photo on p. 108). Then we applied rubber-to-rubber adhesive to both sides of the overlapping membrane (see the bottom photo on p. 108). To keep the splice membrane off the deck membrane while the glue was curing, we put dabs of glue on the top side of the splice and a thin line of adhesive along the membrane on the house wall. These cured quickly and let us temporarily stick the splice out of the way while the glue was applied. Once the splice was glued down, we made diagonal cuts from the outside corners of the door threshold and wrapped the flaps around the door framing and threshold.

Splice is glued to the wall first. A small piece of EPDM had to be spliced in to complete the roof. But because a different adhesive has to be used where the two sheets overlap, the membrane is glued to the wall first.

Rubber-to-rubber glue for the overlap. A special contact cement glues down the overlapping area of the splice. The small dabs on the face helped to stick the membrane up out of the way while the glue was drying.

Roof-Shingle Intersection Gets Special Treatment

Typically, EPDM membranes terminate at the edge of a roof and lap over onto the rake or fascia boards. In these cases, we use a factory-supplied aluminum termination bar to fasten down the membrane edges. A termination bar is a piece of mill-finished aluminum stock $\frac{1}{8}$ in. to $\frac{3}{16}$ in. thick and about an inch wide that is fastened down with screws through predrilled holes (about 12 in. o.c.) over the edge of the membrane to prevent it from lifting in the wind. Roof cement seals the termination bar to the membrane and the membrane to the rake or fascia.

On this rubber roof, the membrane extended over the shingles. Because it's difficult to get EPDM to adhere to roof shingles and because a termination bar didn't seem like an appropriate finishing touch, we used a different detail to end the membrane, known as babing (see the drawing on p. 103). We bent a length of flat aluminum coil trim stock about a foot wide to the same angle as the intersection of the rubber roof and the shingles and then aligned the bend over the intersection. We bedded the trim stock in roof cement and attached it by driving screws every foot or so along both edges.

With the aluminum trim stock fastened down, we trimmed the excess rubber membrane by running a razor knife along the edge of the aluminum, taking care not to cut the shingles. We then glued down the last piece of cured membrane on top of the aluminum, lapping back onto the bottom layer of rubber roofing and creating an attractive transition onto the shingles. On this job, we let this piece of membrane also run up and wrap around the ends of the kneewall.

A peel-off backing keeps the membrane clean. A protective backing peels off the uncured membrane just before adhesive is applied.

Uncured Rubber Flashes the Corners

With all the cured membrane installed, we used uncured rubber to cover and seal the EPDM we cut to fit various corners. Uncured rubber comes in rolls with widths of 6 in. to 2 ft. Uncured rubber is flexible and can be stretched easily to conform to different contours. It comes with a backing that gets peeled off just before the adhesive is applied (see the photo above).

We cut lengths that extended a few inches longer than the areas we wanted to cover. We then trimmed all corners round, which prevented edges of the uncured strip from snagging and lifting up after it was glued in place. We glued down the uncured rubber strips with the rubber-to-rubber adhesive we used for splices in the roof-deck membrane.

As we did with the membrane, we applied adhesive to both surfaces. We first peeled the backing off the strip, then laid it with the peeled side up on a small piece of plywood or OSB that let us spread the adhesive

Detailing Corners With EPDM Membrane

Each time three or more surfaces converge on a rubber roof, the membrane must be cut to conform. After the cut membrane is adhered, strips of flexible uncured membrane are stretched to cover the cuts at the intersecting surfaces.

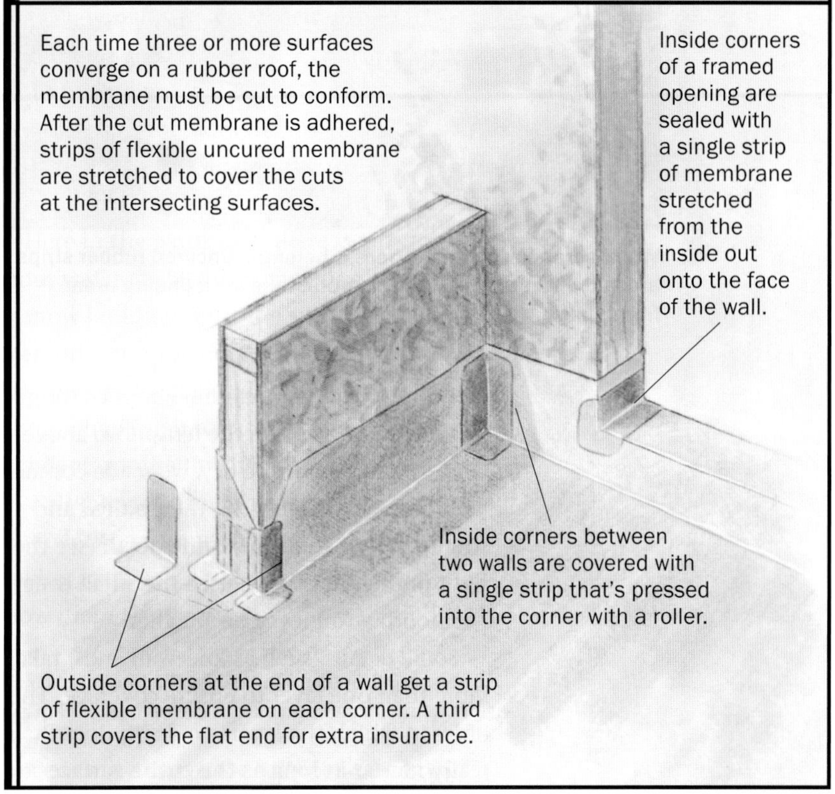

Inside corners of a framed opening are sealed with a single strip of membrane stretched from the inside out onto the face of the wall.
Inside corners between two walls are covered with a single strip that's pressed into the corner with a roller.
Outside corners at the end of a wall get a strip of flexible membrane on each corner. A third strip covers the flat end for extra insurance.

Installing Steel Roofing

■ BY JOHN LA TORRE JR.

During the dry season here in the foothills of the Sierra Nevada, the forests become tinderboxes waiting for a wayward bolt of lightning or a spark from a lighted match. The resulting wildfires create a unique terror for local residents, as the fires leave decimated acres of woodland as well as charred subdivisions in their wake.

So when Ben and Sandy Smith asked me to build a garage next to their house, the first thing that I pointed out was that trees had to be removed for safe clearance from fire danger. Ben responded, "The trees are staying. We're going to build a fireproof garage with a steel roof."

An Argument for Steel Roofing

In addition to its fire resistance, metal roofing is long-lasting, lightweight, easy to install, and easy to maintain. The metal roofing that I installed on Ben's garage is guaranteed to last for the entire life of the building. Only the very best asphalt shingle carries such a warranty.

Asphalt shingles also degrade when exposed to sun, wind, and carpenters' feet. The painted finish on a metal roof will not break down from exposure to weather. And because metal roofing is one solid panel from eaves to peak and is screwed to the roof sheathing, it's not likely to blow off in a gale as asphalt shingles often do.

The tile roofing common in these parts is resistant to fire as well as weather, but its rough surface can collect leaves and debris, especially in roof valleys. Removing debris from a tile roof is complicated by the fact that walking on the tiles can damage them. The maker of this steel roof assures us that it's okay to walk on the roofing. Also, the smooth painted surface of steel roofing discourages debris from accumulating. And when properly installed, it's just plain tough to beat the crisp, clean, colorful lines of a steel roof.

To give you an idea of how the cost of steel roofing stacks up, let's compare the costs of three different types of roofing. In my area, asphalt-shingle roofs are by far the most common and up-front the most economical to install. A 25-year shingle roof costs about $140* per square ($80 for materials and $60 labor). Tile roofs are the most expensive, and although cost varies according to the choice of tile and the shape of the roof, an average tile roof costs about $500 per square ($300 for materials and $200 labor).

Manufacturers of steel roofing recommend that it be installed over plywood sheathing (OSB sheathing doesn't offer enough holding strength for the screws).

Steel roofing falls somewhere in between those two choices. Roofing materials (roof panels and trim) cost around $200, and labor is generally around $150, or about $350 per square total for steel roofing. But when you consider the longevity of a steel roof over asphalt shingles, steel becomes the economical choice.

Be Sure to Order the Right Lengths

One of the trickiest parts of installing a steel roof is ordering the materials correctly. The factory precuts each piece according to your order (see the photo below), so you must first figure out exactly how many pieces you'll need and how long each piece must be. Working off the building plan, I first sketch in each piece of roofing.

The standing-seam roofing that we were using for Ben's garage comes in panels 16 in. wide. With a total roof length of 40 ft., we needed 30 panels (at two different lengths) to cover the front of the roof. The back of the roof was divided into two parts. The first section was 24 ft. 8 in. long, which required 19 panels, and the other section, at 17 ft. 4 in., required 13 panels. The back side of the roof required two more panels than the front side because one section of the back roof overhangs the other.

I measure the panel lengths directly from the roof. Because this roof was to include a ridge vent, the plywood roof sheathing was held 2 in. back from the ridge. To figure the length of each section, I measure from the top edge of the sheathing (where the roofing will end) to the bottom edge and then add an inch so that the roofing extends over the gutter.

Trim and flashing are available to match the color of the roofing. For this project, we needed four different types of trim: sidewall flashing where the side edge of the roof ends at a vertical wall; end-wall flashing where the top end of a shed roof meets a vertical wall; rake trim or gable flashing for the gables; and vented ridge flashing.

Each type of trim and flashing must be anticipated and included in the order. Trim pieces typically come in 10½-ft. lengths, but just adding the total footage and dividing by 10½ might force you to use short pieces to finish a run. Instead, I order the number of 10½-ft. pieces needed to complete each separate run.

The roofing order should also include painted self-tapping screws with rubber gaskets. For standing-seam roofing, painted screws are used for the trim and wherever a screw has to be left exposed to the weather. The screws that secure the roofing panels are hidden, so I use 1-in. truss-head screws that are available at my local hardware store.

Let the factory do the cutting. Steel roofing comes precut to all the lengths you'll need. Here, the different lengths are laid out awaiting installation on the different-size sections of the roof.

Steel Roofing Requires No Special Prep Work

Manufacturers of steel roofing recommend that it be installed over plywood sheathing (OSB sheathing doesn't offer enough holding strength for the screws). We used ⅝-in. plywood for Ben's garage roof.

We then covered the sheathing with 30-lb. felt paper as recommended by the roofing manufacturer. The gutters on the garage were also installed before the roofing. At the roof peak, we installed a length of L-shaped galvanized metal along the top edge of the sheathing as a baffle for any rain that might blow in under the vented ridge flashing. This strip is mostly hidden by the ridge cover, but I painted it blue anyway because I'm fussy.

Where the sheathing was held back from the peak to allow for airflow, I installed a metal insect screen. Metal-roofing companies sell perforated steel for this particular application, but the insect screening is a lot less expensive.

Panels Must Be Perfectly Spaced and Square to the Bottom Edge

To keep the bottom edge of the roof perfectly straight and even, the panels must be installed square to the bottom edge. I begin by marking a large 3-4-5 triangle from the bottom edge of the sheathing to check the roof for square. If it's square, I pull the layout marks directly from the edge of the rake.

If the roof isn't perfectly square, I install the first panel parallel to my square line, making sure the first rib does not hang over the gable edge of the roof sheathing. (Any overhang can prevent the gable trim from fitting tight against the rake.) In this case, the roof was square, and I pulled the layout marks from the edge of the sheathing.

Marking the layout properly is a crucial step to a successful installation. Measuring from the edge of the roof at the peak, I make my first mark at 17½ in., which is the total width of one panel, including the screw tab. From there, I make marks every 16 in. all the way across the roof using a soapstone marker (available at welders' supply stores)

Panel layout is marked along the top and bottom of the roof. Marks are made every 16 in. along the top and bottom edges of each roof section. The edge of each panel is lined up with the marks before being screwed down.

that shows up well on the black felt paper (see the photo above). I duplicate my measurements and make a second set of marks along the bottom edge of the roof as well.

Screw First, Snap Later

We line up the first panel on its top and bottom marks and then screw the panel in place. Special care has to be taken not to overdrive the screws. The screw flange is slotted to allow for slight panel movement

Interlocking ribs snap together. The rib on one side of the panel overlaps and snaps onto the rib of the adjacent panel, hiding the fasteners in the process. Exposed fasteners on the edge of the roof have a rubber grommet to make them weatherproof.

Panels are screwed down before ribs are snapped together. Each new panel is first lined up on the layout marks, and screws are driven at the top and bottom. After the entire flange is screwed down, the seams can be snapped together with gentle pressure from the palm of your hand or the sole of your shoe.

during normal expansion and contraction. The screws should be snugged against the flange, but not so snug that the flange deflects under the screw head.

Each panel connects to its neighbor via overlapping ribs that snap together (see the photo at left), and the temptation is to snap the ribs together before you screw down the panel. But I've found that some of the panels get slightly stretched or compressed in shipping. So if you snap the ribs together first, you may not be able to push or pull the screw flange to its proper location. Unless you follow the layout exactly, you can gain or lose up to ⅛ in. per panel. Over the course of a 40-ft. roof, this discrepancy could add up to almost 4 in.

Instead, the bottom edge of each panel is lined up with its neighbor. The screw flange is then set at the marks and screwed down before the ribs are snapped together (see the photo below left). Because installation goes pretty quickly, we often screw down two or three panels, and then one crew member goes back and snaps the ribs together with gentle shoe or palm pressure.

When we reach the other end of the roof, we again make sure that the last panel does not stick out past the edge of the barge rafter. If we don't end with a rib along the edge (as was the case with one of the back sections of Ben's roof), we measure the remaining distance, add an inch and cut the panel to that width. The extra inch of material is bent up with a hand seamer (Malco Products; 800-596-3494) to form a rib (see the top left photo on the facing page).

There are a number of ways to cut sheet metal quickly and accurately. I own an electric nibbler that munches its way through metal roofing, following almost any pattern I want (see the top right photo on the facing page). I try to do my nibbling over a trash barrel to catch the tiny metal shavings that the machine produces.

But the best way I've found to do straight-line cutting is with dueling tin snips (see the bottom photo on the facing page). When

Special pliers help to bend the roofing. To make a rib for the edge of a roofing section, special pliers called hand seamers bend the edge in gradual increments.

Munching metal. A tool called a nibbler can cut just about any shape in a metal-roofing panel, but the shavings are messy.

installing metal roofing, I keep pairs of right-handed and left-handed snips in my tool belt at all times (and yes, for the nautically inclined, the left-handed snips have red handles, and the right-handed have green).

The trick is using one pair in each hand at the same time. The first pair of snips follows the cutline, and the other makes a parallel cut about an inch away. The waste curls up safely and easily between the two snips. I'm able to do the cutting right there on the roof, and there are no metal shavings or extra power tools to contend with.

Special Details for Roof Penetrations

Anything that goes through a metal roof is a potential cause for a leak. Small penetrations such as plumbing vents or gas vents are sealed with special flashings called roof jacks made specifically for metal roofing (see the left photo on p. 118). These roof jacks consist of a conical boot and a flat flange made of soft, flexible rubber.

Dueling snips. To cut straight lines in a roofing panel, right-handed and left-handed snips are used side by side, and the waste curls up harmlessly in between.

Rubber roof jack seals smaller pipes. Small round roof penetrations such as gas-vent pipes are sealed with a rubber roof jack that is caulked and screwed to the roofing.

When a chimney falls on a seam. When a large penetration such as a chimney is in the middle of a rib, a special rib is built into the lower apron of the flashing (above). Roofing sections are then cut, fit and screwed over the upper part of the flashing (below).

The ribs on a standing-seam roof are too abrupt to mold around the roof jacks, so vents must fall between ribs, which is easy to plan for with new construction such as Ben's garage. First, caulking is applied in a double bead to the bottom of the flange. Then the jack is pushed down over the pipe until the flange contacts the roof. Next, I drive gasketed screws every inch or so around the perimeter of the flange.

Large penetrations such as chimneys need a different treatment. This roof had two metal fireplace chimneys and an evaporative cooler. The cooler spanned two roof panels, so I turned to my trusty sheet-metal man, Dave Doyle, to make a custom roof jack of flat 26-ga. galvanized steel.

Dave made the cooler jack as one piece, and the roofing panels were cut around it. The bottom jack apron has a raised rib that covers the standing seam of the panels below. The top apron was kept flat so that the roofing panels could lie on top of the jack.

Each chimney required special attention as well. The first chimney fell smack dab in the middle of a rib. But Dave's artistry made the solution simple. The roof jack supplied by the chimney company could not be modified, so Dave took a roof jack made for a gas vent and welded a raised rib on the bottom apron (see the top right photo above). Again, the top apron was left flat, and I just had to cut and screw down the roofing panels around the chimney (see the bottom photo above).

The second chimney fell between ribs, but this time, the roof jack from the chimney company was too large. Fortunately, the same size roof jack that Dave had modified for the first chimney fit beautifully between the ribs. I cut the apron, leaving extra material on each side that I bent up to tie in with the roofing ribs (see the top left photo on the facing page). I cut the roofing panel so

When a chimney falls between seams. If the penetration is between seams, the edges of the chimney flashing are bent up to form ribs and installed over the lower-panel section (left). An upper-panel section then fits over the flashing and is screwed into place (below).

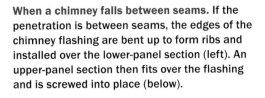

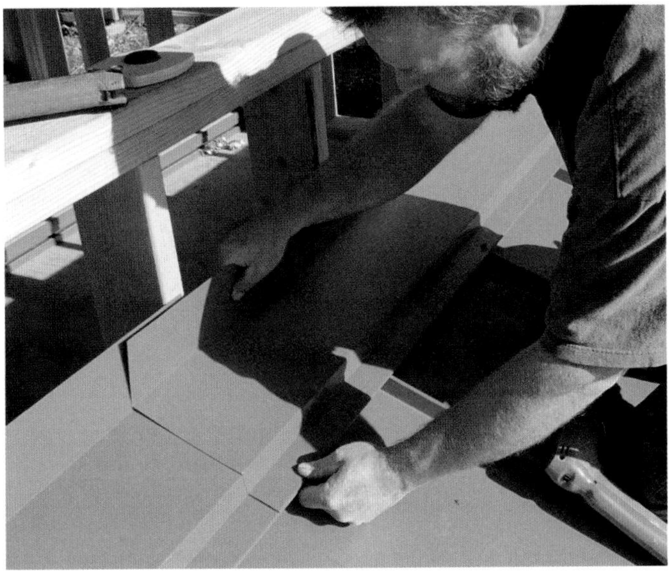

Overlap trim pieces in long runs. When a run of trim requires two or more pieces of trim, the pieces of metal are overlapped about 6 in. and then held in place with gasketed screws.

that it ended just short of the chimney. The jack then overlapped that piece, and a top panel section overlapped the flashing and wrapped around the chimney (see the right photo above).

By the way, cutting through a rib with snips inevitably crushes the rib. To bang it back out, turn the panel upside down and open up the rib with a nail set.

Finishing the Trim: Origami in Steel

Screwing down the roof panels always goes quickly. Installing the trim is the more time-consuming part.

On runs of more than 10½ ft. that require more than one length of trim, I overlap the pieces by 6 in. or so (see the bottom left photo above). The material is thin enough that the overlaps are not noticeable. Trim is attached with gasketed screws; again, I take care to drive the screws enough to flatten the rubber washer but not enough to deflect the roofing or the trim.

The tricky part is finishing the ends of each trim run. Most varieties of trim have an open space when they are viewed from the end. Not all steel-roofing installers go to the trouble of shaping the ends of the trim to close these spaces, and I've seen metal roofs with openings that were big enough to

Tidying Up the Ends of the Trim

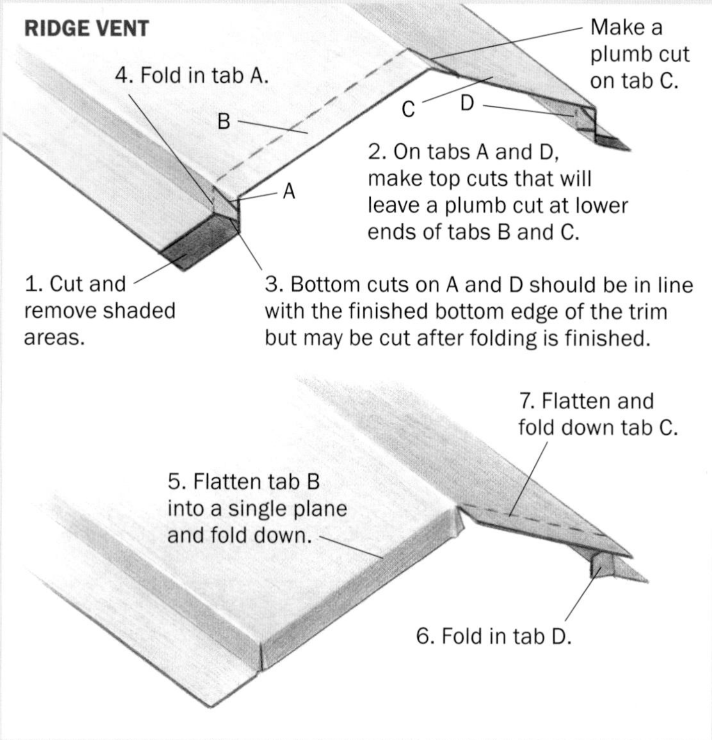
RIDGE VENT
4. Fold in tab A.
B
C
D
Make a plumb cut on tab C.
A
2. On tabs A and D, make top cuts that will leave a plumb cut at lower ends of tabs B and C.
1. Cut and remove shaded areas.
3. Bottom cuts on A and D should be in line with the finished bottom edge of the trim but may be cut after folding is finished.
7. Flatten and fold down tab C.
5. Flatten tab B into a single plane and fold down.
6. Fold in tab D.

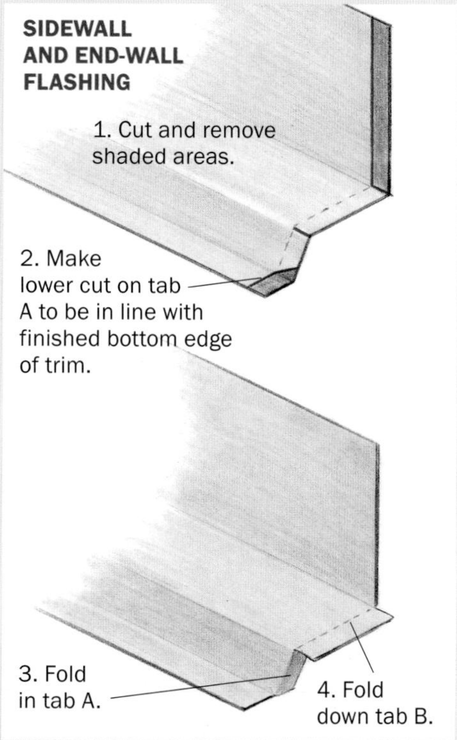
SIDEWALL AND END-WALL FLASHING
1. Cut and remove shaded areas.
2. Make lower cut on tab A to be in line with finished bottom edge of trim.
3. Fold in tab A.
4. Fold down tab B.

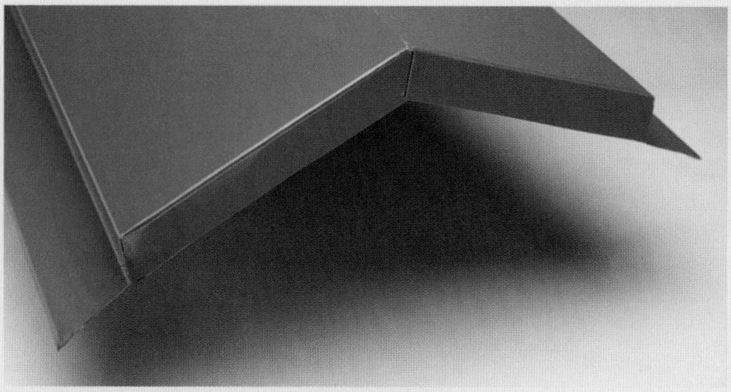

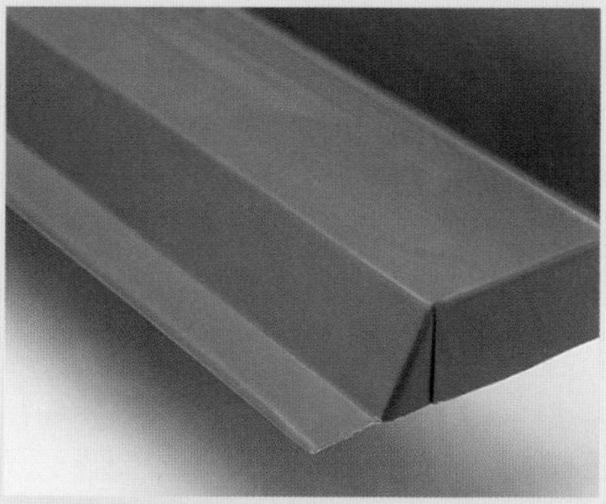

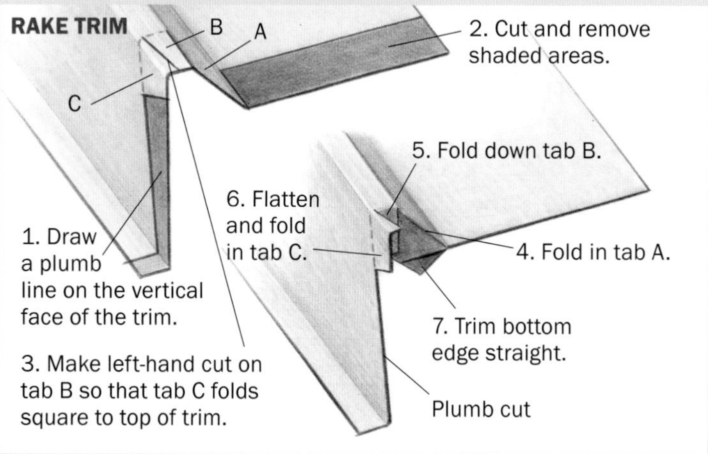
RAKE TRIM
B
A
C
2. Cut and remove shaded areas.
5. Fold down tab B.
6. Flatten and fold in tab C.
4. Fold in tab A.
1. Draw a plumb line on the vertical face of the trim.
7. Trim bottom edge straight.
3. Make left-hand cut on tab B so that tab C folds square to top of trim.
Plumb cut

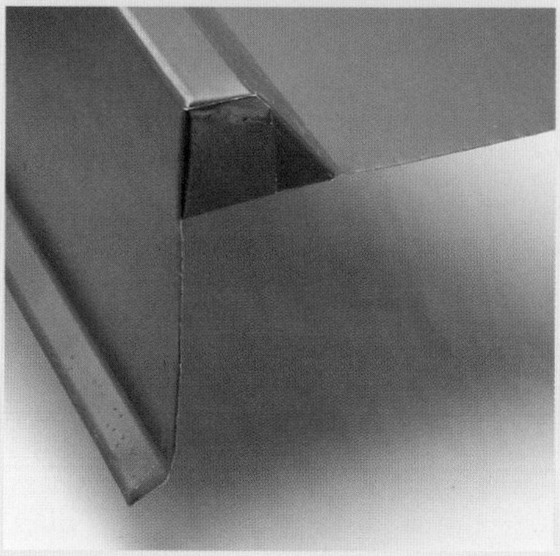

Chalklines guide ridge-vent installation. Marks are made on the roofing from a scrap of ridge vent, and lines are snapped between the marks (far left). The vent is held to the chalkline and screwed down to keep it running in a straight line (left).

Screws keep the bottom edge flat. Gasketed screws are driven between the ribs of each panel along the bottom edge to keep the panel flat.

Sources

Steel roofing can be divided into two basic types, through fastener and standing seam. A through-fastener roof is installed with prepainted screws driven through the ribs.

Standing-seam roofing, as was used on this project, has interlocking vertical seams. The fasteners are hidden beneath the seams for a cleaner profile and less chance of rain getting in at the fasteners. But the trim is still fastened with exposed screws, reducing this advantage.

Standing-seam roofing typically costs about 25% more than through-fastener roofing; the labor and trim cost about the same. Steel roofing is available in a wide variety of colors.

The steel roofing for this project was made by ASC Profiles® (800-360-2477; www.ascprofiles.com), but here are some other companies that make metal roofing.

AEP-Span®
(800) 527-2503
www.aep-span.com

Atas International
(800) 468-1441
www.atas.com

Fabral®
(800) 477-2741
www.fabral.com

Follansbee® Steel
(800) 624-6906
www.follansbeeroofing.com

McElroy Metal
(800) 950-6531
www.mcelroymetal.com

Metal Sales
(800) 406-7387
www.mtlsales.com

Morin Corp.
(800) 640-9501
www.morincorp.com

Nu-Ray Metals
(800) 700-7228
www.nuraymetals.com

Wheeling-Pittsburgh Steel Corp.
(877) 333-0900
www.wpsc.com

throw a cat through. It may take a bit more time, but I prefer to cut and fold the ends of the trim to give the roof a more finished look (see the sidebar on the facing page).

Rake trim should be installed from the bottom of the roof, working up to the peak with each upper piece overlapping the one below. For the ridge vent, I place a short section on the peak so that it lies evenly side to side. I mark the outside edges, and then I repeat the process at the other end. Chalklines are snapped between the marks (see the top left photo above), and the ridge vent is set on these lines as it's installed (see the right photo above).

When we've finished with the trim, the final step is driving a couple of gasketed screws along the bottom edge of each panel (see the bottom left photo above). The only recommended maintenance is an annual washing with clean water. And if this year is any indication, I'm pretty sure Mother Nature will take care of that.

**Prices noted are from 2000.*

John La Torre Jr. is a carpenter in Tuolumne, California.

Choosing Roofing

■ BY JEFFERSON KOLLE

A lot of roofing materials try to look like something else. When they first came out, asphalt shingles were touted as looking like slate. Today, there are metal roofs that are supposed to look like tile, and there are fiber-cement roofs that are supposed to look like wood shingles.

Whether you're looking for a roof that looks like something else or a roof that looks like what it is—wood roofs really do look like wood—there are a lot of materials on the market, and they all have their benefits. Material costs and installation costs of some roofing materials are higher than others, but the payback is in their longevity or in their aesthetic appeal. What follows is a survey of the most common roofing materials available for steep-roof residential construction (anything greater than a 3-in-12 pitch).

The standard unit of measurement for roofing materials is the square. A square is 100 sq. ft. of roofing. Manufacturers refer to their products on a per-square basis—cost per square, weight per square, etc. This article will use the same nomenclature.

Asphalt Roofing Is Inexpensive and Can Be Installed Quickly

One story has it that the three-tab asphalt strip shingle, with its two grooves dividing the exposed face of the shingle into three sections, was invented by Fred Overbury in 1915 when he pulled a cardboard divider out of an egg crate and was struck with a brilliant idea. Before Overbury's invention, asphalt shingles were made as individual pieces and were installed one at a time. The strip shingle revolutionized the asphalt-roofing industry.

Today, asphalt shingles cover more residential roofs than any other material. Every year, 100 million squares of asphalt shingles are installed in the United States. That's more than 358 square miles, about the size of Lake Tahoe.

Asphalt shingles' popularity is due to several factors. They're fire resistant, and choices of color and textures are numerous. They're relatively inexpensive both to purchase and to install; an experienced roofer can install 10 to 20 squares a day, depending on the intricacies of the roof.

Architectural shingles cost more but last longer. Architectural asphalt shingles cost twice as much as three-tab shingles. The advantages are a longer warranty and the expectation of a longer life.

On a sunny summer day, a black asphalt roof can reach a temperature of 150°F. As the temperature rises, asphalt shingles become soft and pliable. A sudden thunderstorm can cause the temperature of that soft, pliable roof to drop to around 60°F. That's called thermal cycling. If the asphalt shingles on the roof are going to continue to shed water, their reaction to thermal cycling has to be minimal. Of course they will shrink when the temperature drops. But they can't curl, and they can't lift off the shingle below.

All asphalt shingles share common construction: A reinforcing mat is impregnated with asphalt. Twenty-five years ago, fiberglass mats were introduced to replace the earlier organic-fiber mats. Although organic-fiber mat shingles still are recommended for areas with extreme winds, early blow-off problems with fiberglass-mat shingles have been eliminated so that, today, fiberglass-mat shingles are the most common type sold.

Filler materials, most commonly ground limestone, help stabilize the shingle's asphalt—technically a liquid—by stiffening it and keeping it from flowing. The fillers' inertness adds to the fire retardancy of the shingles, and they increase resistance to cupping during thermal cycling.

Asphalt degrades in sunlight; it loses its suppleness, dries out, and cracks. To combat that problem, asphalt shingles have a surface coating of granulated minerals pressed

Each year 100 million squares of asphalt shingles are used in the U.S. That's an area about the size of Lake Tahoe.

COMPARISON OF ROOFING MATERIALS

	Asphalt	Wood	Metal	Tile	Slats
Cost/square	$25–$56	$150–$200	$35–$250	$120–$1,000	$350–$700
Installation* cost/square	$65–$125	$130–$160	$35–$400	$100–$300	$250–$450
Approx.** life span/yrs.	15–20	10–40	15–40+	20+	30–100
Weight in lb.	225–385	300–400	50–270	375–1,100	500–1,000
Fire rating	A	B***	A	A	A

*Installation costs vary enormously due to many factors, such as local labor rates, time of year, complexity of a house's roof geometry, height of a roof from the ground, and complexity of a roofing material's profile.
**Roofing materials' life spans are courtesy of the American Society of Home Inspectors.
***Wood shingles and shakes treated with a fire retardant are given a Class B fire rating. Untreated shakes and shingles have no fire rating. A Class A roof is attainable with wood roofing, but a special installation procedure involving a sheathing sandwich made of plywood and gypsum board is necessary.

into the part of the shingle exposed to the sun. Eventually, when the mineral granules wear off a shingle, through abrasion or erosion, the shingle degrades quickly. The granules are what give a shingle its color. Colors from bright greens to blues, yellows, and reds are available as well as blacks, whites, and a variety of subdued earth tones.

Architectural Shingles Are Thicker Than Three-Tabs

About the time fiberglass reinforcing mats were introduced, manufacturers came out with what are known as architectural or laminated shingles. Unlike three-tab shingles with cutout grooves, architectural shingles typically are solid across their length. Multiple, overlapping layers are laminated to form a heavier shingle with a more textured appearance.

Some architectural-shingle manufacturers use different colored mineral granules on the multiple layers to form the illusion of the cast shadowlines (see the left photo on the facing page) one might see on a wood-shake or slate roof. From the sidewalk, at dusk, in the fog, one might think a cedar-colored architectural-grade asphalt-shingle roof was wood. Other than that, they aren't convincing. The slate imitators also are unconvincing.

Most three-tab shingles weigh around 240 lb. per square, but some architectural shingles can weigh as much as 100 lb. more per square. Three-tab shingles come with a 15-year to 20-year warranty, but, because there is more material in the architectural shingles, they come with a longer warranty, typically 30 to 40 years (see the photo on p. 123). Architectural shingles are sold at a premium. In Newtown, Conn., for instance, three-tab shingles sell for $24* per square; architectural shingles cost $52 per square.

Architectural asphalt shingles try to look like wood. Different colored surface granules and overlapping layers on thick architectural shingles attempt to mimic the textures of a wood-shingle roof.

The look of real wood shingles. The natural beauty of a wood roof palliates its high cost. These pressure-treated red-cedar shingles are warranted for 30 years against fungal decay.

There's a common thought that a heavier shingle is a better shingle. But according to a paper sponsored by the National Roofing Contractors Association, "Shingle testing and observations from field performance have frequently shown that weight alone is not a sufficient indicator of shingle quality…" Rather, "the quality of the individual components of the composite structure—the reinforcement [mat], the asphalt, and the filler—are much better indicators of shingle performance." According to W. Kent Blanchard, one of the authors of the report, there is no easy way for a consumer to get an indication of the quality of a shingle's components, but he said to make sure any fiberglass-mat shingle you purchase has passed ASTM standard D6432, a measurement of tear strength. According to an article in the September 1992 issue of *Roofer Magazine*, "High tear strength is a good indicator of shingle toughness and resistance to cracking."

Wood Roofing Is Beautiful

Regardless of manufacturers' attempts to simulate the appearance of wood shingles, nothing really looks like a wood roof except the real thing (see the right photo above). Although wood roofs are expensive to purchase and to install, many people think their aesthetic value outweighs their high cost.

Wood roofs are composed of either shingles or shakes, and, although both are wedge-shaped in section and often are confused for one another, there are basic differences that set them apart. Put simply, a shingle's tapered shape is attained by sawing, and a shake's is split, or rived, from logs. Shakes are thicker at the butt, or bottom, than shingles, and the striations that result from splitting along the length of the grain give shakes a more rustic and textured appearance than sawn shingles.

Wood roofs are not only expensive to purchase—a square of top-quality shakes can cost more than $150—they also are labor intensive to install. Because wood shakes and shingles are nailed onto a roof one at a time, it takes a while to cover a roof with

Clean keyways add life to wood roofs. Power washing, broom sweeping, or rinsing with a hose will clean rot-causing debris from between shingles. Periodic treatments with wood preservative will help extend the roof's life.

Except for certain metal roofs that can be repainted, wood is one of the few roofing materials that can be maintained.

wood. A fast shingler on a section of roof without any time-consuming obstructions to roof around can put on two or three squares per eight-hour day.

Preparation of the deck before roofing can begin also takes a while. Good practice dictates laying shingles or shakes over skip sheathing, solid boards nailed with spaces between the rows, rather than over plywood or a similar solid decking. Skip sheathing is preferable for a wood roof because the air spaces between boards allow the wood roofing to dry from both sides, thus contributing to the roofing's longevity.

MAINTAINING A WOOD ROOF

Except for certain metal roofs that can be repainted, wood is one of the few roofing materials that can be maintained. And, in fact, it is recommended that wood-roof maintenance will protect an expensive investment. Brian Buchanan is a wood technologist in Lufkin, Tex. He has written a treatise called *Evaluating Various Preservative Treatments and Treating Methods for Western Red Cedar Shingles*. Buchanan told me that the most important practice toward promoting a wood roof's longevity is to keep the keyways between shingles clear of debris (see the photo at left). When the keyways get clogged—with leaves, conifer needles, dead birds, whatever—fungus starts its eventual course toward rot. Power washing, sweeping with a broom, and even simple washing with a garden hose cleans out keyways. Keep in mind that a wet wood roof is a slippery wood roof.

Liquid preservatives for wood roofs vary in their effectiveness. The best contain copper naphthenate. While some might consider the green color of products containing this compound to be unsightly, there are pigmented versions that simulate cedar's reddish-brown color. A quick read of a can's label can tell you if the product contains copper naphthenate. A free copy of Buchanan's pamphlet is available from the Texas Forest Service (409-639-8180).

Western red cedar covers more roofs than any other species. Alaskan yellow cedar shingles are available as are ones made of Eastern white cedar. And, recently, the Southern Pine Council™ (504-443-4464) has begun touting pressure-treated Southern pine shakes.

WOOD ROOFS AND FIRE

Red-cedar shingles that have been impregnated with a fire retardant are given a Class B fire rating (Class A is the most fire resistant). A Class A roof can be had using wood, according to the Cedar Shake & Shingle Bureau, but you have to install the fire resistant shingle over a sandwich of two layers of sheathing with a layer of ½-in. gypsum board between.

Fire retardants work well; the wood won't burn after it is treated. The problem comes from the treated wood's exposure to weather. Rain soaks the wood roof, and, when the sun comes out and dries the wood, the retardants are drawn to the surface. Subsequent rains wash the retardant off the wood. The process is repeated over the years, and, when all the retardant leaches out of the wood, your roof is covered with kindling.

But the Cedar Shake & Shingle Bureau has another view of the situation. Don Meucci, a spokesman at the bureau, said that tests done of fire-treated wood shingles taken from a roof 16 years after installation passed the same stringent tests that new shingles must undergo today.

But communities across the country from Los Angeles, Calif., to Newcastle, N.H., have banned wood roofs, even those treated with fire retardants, because of their flammability. Of course, if a fire starts in your kitchen, no roof is going to keep your house from burning. The problem with wood roofs is twofold: Sparks that land on a wood roof can cause it to burn, and when the wood roof catches fire, the wood can send off flying brands, bits of burning material that leave the roof with the smoke column and then fall to the ground, still glowing hot, ready to start the next fire.

If you're considering a wood roof, talk to the Cedar Shake & Shingle Bureau (206-453-1323), your building department, or your local fire chief.

Metal Roofs Aren't Just for Barns Anymore

My first memory of metal roofing is not a pleasant one. My brothers and I were on my grandfather's farm, and we discovered that the big barn was full of bats. One of us had the brilliant idea to rid the barn of the flying rodents with bow and arrows. We didn't hit any bats, but we did puncture the roof three or four times. My grandfather saw the arrows sticking though the galvanized sheets of roofing. He was not pleased.

Metal roofing is no longer relegated just to farm buildings, and it is available in styles and colors other than the rusting galvanized tin seen everywhere from Walker Evans' early photographs of sharecroppers' houses in Tennessee to my grandfather's barn in southern New Jersey. Improvements in both metal-coating processes and in waterproof fasteners have made metal roofing suitable for residential use. According to Sig Hall, an estimator for Bryant Universal, the largest roofing contractor in the United States, metal roofs are being put on homes in a number that is increasing faster than any other material.

Metal roofs have been around for a long time. First fashioned on site by hand and later manufactured in sheets installed in long sections, metal roofing is available in patterns other than standing seams or sine-wave corrugations. Some manufacturers make panels that are supposed to simulate clay tiles (see the photo below).

A variety of metals is used for roofing—everything from copper to stainless steel to aluminum to alloys and coatings and compounds of each. Painted finishes are available in a range of colors as wide as what is offered on today's cars. Most manufacturers will blend custom colors if you're really picky and can't find something in their stock palette. Because of its thinness, metal roofing is the lightest of all roofing material. There are coated aluminum shingles that weigh only 50 lbs. per square (see the photo on p. 128). Because metal roofs are so light, the metal-roofing industry touts metal as an excellent choice for reroofing. Depending

Metal that tries to look like tile. Painted-metal panels weigh only 125 lbs. per square. Their eaves-to-ridge length makes for fast installation.

Aluminum shingles are light. Coated aluminum shingles weigh only 50 lbs. per square. Touted as looking like wood, their textured appearance is unique.

on the application—some materials might require the addition of furring strips or standoffs—a metal roof can go over an existing asphalt or wood roof.

The metal-roofing folks also harp on their product's ecological benefits. Because no tearoff is required, no shingles, asphalt, or wood go into landfills. Another benefit to the environment is that metal roofing is the only completely recyclable roofing product. When a metal roof does wear out, and most metal roof are expected to last 50 years, it can be recycled into new roofing. And there's a good chance that the metal roof you put on your house used to be a beer can.

Some manufacturers make metal shingles, but most metal roofing comes in panels that go on quickly. Because of the long eaves-to-ridge panels that are available—some manufacturers have shipped panels 80 ft. long—a lot of square footage gets covered at a time. Depending on the complexity of a roof's geometry and the complexity of the profile of the metal roofing being used, an experienced installer can lay down between 3 and 30 squares in a day.

The cost of a metal roof depends on the kind of metal used, the thickness of the material, and the finish. According to Rob Haddock, the director of the Metal Roof Advisory Group, the least-expensive metal roof—galvanized tin sheets—can be purchased and installed for as little as $60 per square, while handcrafted copper on a complex roof could cost as much as $1,500 per square. In an article in *Contractors Roofing and Building Insulation Guide*, Haddock wrote that "metal roofing is the lowest cost, highest cost and everything-in-between roofing material."

Standing-seam roofs are the most popular profile (see the photo on the facing page); they have a vertical seam that stands proud of a flat panel. Panels are joined together at the edge seams either by a crimping of the seams or by a cap that covers them. A standing-seam panel can be formed either at the factory or on site by a portable roll-forming machine that bends a coil of metal.

If you live in snow country, keep in mind that metal roofs, especially eaves-to-ridge panels, shed snow quickly, kind of like sledding in reverse. Snow cover on a metal roof will have a tendency to let go: That is, it will slide off the roof all at once, like a miniature avalanche. It might be a good idea to keep your foundation plantings 3 ft. or 4 ft. away from the house if you don't want to lose that prize pyracantha to snow damage. Slick metal roofs also shed rain faster than some other, more textured roofing materials. Because of this, gutters need to be sized accordingly or, if possible, eliminated.

Tile Roofing

Most people think of tile roofs as indigenous to Florida and the Southwestern part of the United States and as appropriate only to Spanish-style or Mediterranean-style houses. But tile roofs are popular in Europe. And in Japan, tile's popularity rivals that of asphalt in this country. In Japan there are 500 companies making roofing tiles. In this country there are five.

Standing seam can look great on a home. The vertical ribs of a standing-seam roof give the metal panels rigidity. The slick surfaces shed water and snow quickly. Because of this, gutters should be sized accordingly or, if possible, eliminated.

Tile roofs have a textured look. Flat tiles are available, but the soft undulations of a barrel-tile roof or the crested-wave appearance of a Japanese tile roof are markedly more textural than the appearance of most other roofing materials, which, aside from the slight deviation of course lines or seams, are planar in appearance.

Whereas most other roofing materials are classified by the dominant material in their composition—wood, asphalt, metal—roof tiles are referred to as such because of the process of their manufacture. Like interior house tiles, roof tiles are made of a soft, plastic material—either clay, concrete, or fiber cement—that is molded or extruded and then hardened into a brittle, inert state either by heat or by chemical reaction.

Because of its inertness, tile won't burn. You'd have the same difficulty getting a roof tile to burn as you would if you tried to ignite a concrete block or a clay flowerpot. All roof tiles have a Class A fire rating. After the 1992 fires in southern California, some of the only houses left standing had tile roofs. No house is immune from fire, but sparks on a tile roof won't cause immolation.

The question of weight always arises when people talk about tile roofs. There are glazed-clay tiles that weigh more than 1,100 lb. per square, but keep this in mind: Three layers of asphalt shingles (an original and two reroofs) can weigh about 900 lb. And all new-house roofs are engineered to carry three asphalt roofs. On the other hand, there are lightweight concrete tiles that weigh as little as 375 lb. per square. If you are considering a tile roof, and if you have any questions about your house's ability to withstand the weight, talk to an engineer. He can assuage your doubts or, possibly, suggest some roof reinforcement that might not be as expensive as you think.

Are roof tiles expensive? Well, how much does a new car cost? Both questions need qualification before an accurate answer can be given. According to Stu Matthews, owner of Northern Roof Tile Sales in Ontario,

Because roof tiles are heavy and because there are a limited number of manufacturers in this country, shipping costs to a job site can have a huge effect on the cost of a tile roof.

Canada, the price of a roof tile is not an accurate indication of its quality. The quality of barrel tile costing $600 per square is similar to tiles with the same profile that cost $120 per square.

If price does not indicate quality, what does affect the price of tile? Matthews said several things affect cost. As in most commodities, a manufacturer's production volume affects the price at which it can offer its goods. A company with automated manufacturing procedures, making 150 million tiles per year, can sell its tiles for a lot less than a manufacturer making 1⁄10 that number.

What else affects cost? There's a British saying: "You can tell a person's wealth by the size of the tiles on his roof." Contrary to the popular notion that bigger is always better and more expensive, in the case of roof tiles, the opposite is true. On a per-square basis, smaller tiles cost more to purchase and to install. The smaller the tile, the more tiles there are per square, and, because tiles are installed one at a time, smaller tiles are labor intensive to put on a roof. The most expensive tile that Matthews sells is handmade, and there are 500 tiles per square. They sell for about $1,000 per square. As a point of comparison, there are terra-cotta roof tiles, available in California where they are made, that have 75 tiles per square and cost about $75 per square.

Because roof tiles are heavy and because there are a limited number of manufacturers in this country, shipping costs to a job site can have a huge effect on the cost of a tile roof. The same $75-per-square tile mentioned above could double in price by the time it reaches a job in Massachusetts.

Clay, Concrete, or Fiber Cement?

Unlike interior tiles, which are made, for the most part, of a natural ceramic material such as clay or porcelain, roof tiles can be made of concrete or fiber cement. All materials have distinct advantages.

Clay was the first material used for roof tiles, and, in this country, it is still the most popular (see the photo below). Terra-cotta tiles are flower-pot color and commonly are used on a roof in their natural color. Terra-cotta can be colored by different methods. Engobe is a process in which a colored wash is put on tiles. When raw tiles are fired in the kiln, they take on the color of the wash. Engobe-fired tiles are limited to muted earth tones. Glazing is another process altogether. After an initial firing, tiles are coated with a glaze and fired again. Bright, primary colors are standard offerings from manufacturers that sell glazed tiles, and a lot of companies make custom colors on request. Also, clay tiles won't fade in the sun like concrete tiles.

The look of a real tile roof. The color of clay-barrel tiles won't fade in the sun.

A final note on ceramic tiles: Celadon® Ludowici® (4757 Tile Plant Rd., New Lexington, OH 43764; 800-917-8998) is an interlocking ceramic tile that looks like slate (see the bottom photo at right), but not in the way that some asphalt shingles are supposed to look like wood. Rather, these ceramic tiles have, to my eye, a genuine slate appearance.

Concrete tiles are less expensive than real clay tiles; some cost as little as $50 per square. They are available in profiles that simulate clay-roof tiles. Imitations of slate, wood shakes and wood shingles are all available in concrete. Concrete tiles are heavy—around 900 lbs. per square—but Westile® (8311 W. Carder Court, Littleton, CO 80125; 800-433-8453) makes a concrete tile with the paradoxical name FeatherStone® that weighs 660 lb. per square.

Concrete can be colored, but air pollution fades colored concrete. Matthews recommended using concrete tiles that have the color mixed through the concrete, rather than ones that have a wash of pigment applied to the surface.

Fiber-cement tiles have been around for a long time. Unfortunately, the fiber in fiber-cement tiles used to be asbestos. Manufacturers are vague about what replaced asbestos in fiber-cement tiles, but the new ones contain no asbestos. Fiber-cement roofing tiles are a lot lighter than either concrete or clay tiles. Fiber-cement tiles cost between $225 and $275 per square, and they are made to imitate clay tiles, slates, wood shakes, and wood shingles.

With the ban on wood roofs in many parts of the country, fiber-cement roof tiles that imitate wood roofs are catching on (instead of catching fire). Re-Con Building Products, Inc. (P. O. Box 5659, Eugene, OR 97405) has a fiber-cement roofing product that imitates wood shakes and shingles. The name of the product? FireFree Plus®.

In areas outside Florida and the Southwest, tiles are beginning to become popular. If a roofer tries to convince you that you don't want a tile roof, ask him how many he's installed. Chances are, his experience is limited. Some tiles are more difficult to install than others, but the hardest roof you'll ever tile will be the first one. Depending on the tile—the installation of some are more labor-intensive than others—an experienced roofer should be able to put on between two and three squares per day. Tile manufacturers are excellent at disseminating installation literature.

One is slate, the other is molded clay. The photo at the top is of a real slate roof, and the photo on the bottom is of a Celadon clay-tile roof. Or is it the other way around?

Slate roofs can last indefinitely. The fasteners and flashing will wear out before the roof slates will. Because of slate's long life, there is a thriving market in used slate.

Slate Roofing Can Last Hundreds of Years

In London, there's a building called Westminster Hall that was finished at the turn of the 10th century. They put a new roof on the place in the 13th century, and a smart contractor chose slate. The same roof is still on the building.

Roof slates should never wear out (see the photo above). Given that the material is already a couple of million years old, expecting it to last another hundred years or so on your roof isn't really asking a lot. What do fail are the fasteners that hold the slates to the roof and the flashing at junctures such as valleys and around chimneys.

Because of its longevity, slate is the only roofing material that sometimes can be purchased used. Slates carefully are removed from a roof, the cracked or damaged ones are culled out, and the remaining slate can be reinstalled on another building. Can you imagine putting used asphalt or used wood shakes on a roof?

Slate is expensive. A square of slates can cost between $350 and $700. Add to that the expense of installation, between $250 and $450 per square, according to Terry Smiley, a slate roofer in Denver, Colo., and you've got an expensive roof. But divide the per-square price by 100 years, and it doesn't seem so expensive.

A slate roof might not be as heavy as you think. Depending on the slate's thickness—3⁄16 in. is the industry standard (see the bottom photo on p. 131)—a slate roof can weigh between 650 lb. and 1,000 lb. per square.

Care must be taken when nailing slates on a roof. Sort of the Goldilocks syndrome: not too hard, not too soft. Slates have to be nailed just right. Nailing a slate too tightly will cause it to crack when the slate expands and contracts. And nailing a slate too loosely can cause the slate above to crack. Bill Markcrow at Vermont Structural Slate in Fair Haven, Vt., favors slate hangers. Hangers have been around for a hundred years, but they had fallen from popularity. Markcrow thinks slate hangers are foolproof. They have a hook on one end and a nail on the other. Using them is easy; you still have to nail the slates around the perimeter of the roof, but, for the rest, you just nail on the hangers and place the slates on the hook. The hangers are made of stainless steel, and you can get them painted black. Markcrow says the hangers are visible hooked under the bottom of the slate, but "from 8m away, they disappear."

It's not likely that you'll be able to go to your local roofing-supply store and take home enough slates to roof your house. Slate will have to be ordered from a quarry that will cut it from the ground and then fabricate the material into roof slates. Expect a month or so between the time you place an order and get delivery. But in the larger time frame of a slate roof's life, what's another month?

**Prices noted are from 1995.*

Jefferson Kolle is a senior editor at The Taunton Press, Inc.

Roofing With Slate

■ BY TERRY A. SMILEY

I worked on my first slate roof when I was 14 years old. The farmer whose house we were remodeling also had a huge barn, and several roof slates were missing. He asked my grandfather if we could replace the slates, and because I was the lightest and the most agile, I was recruited. Ten years later, a customer I was installing a slate floor for asked if I'd be interested in installing a slate roof. By then, I'd worked with a lot of slate and was ready to give a whole roof a try. I called the supplier from whom I'd gotten flooring slate, and he sent me a copy of *Slate Roofs*, first published by the now-defunct National Slate Association in 1926. Everything worked out fine, and since then I've put on many slate roofs from Pennsylvania to Arizona. (*Slate Roofs* has been reprinted by Vermont Structural Slate Co. Inc., P. O. Box 98, Fair Haven, VT 05743; 800-343-1900. The book sells for $11.95.)

Slate is the ideal roofing material. It won't corrode or burn, and for the most part it won't wear out. I'm not saying that a slate roof will never need replacement or repair. But it is important to keep in mind that a roof is a system of its materials, and a slate roof is made up of more than just pieces of sedimentary rock pulled from a quarry, split by hand and cut into rectangles. The integrity

Use good materials for a long-lasting roof. The author uses high-quality materials such as copper flashings and fasteners, bituminous membrane, 30-lb. felt, and silicone caulk to ensure his roofs last as long as the slate. Wood strips run up all hips and across the ridge to provide nailing for the saddle caps applied last.

Special tools for roofing with slate. A slate ripper, on the left, is used to cut or pull nails when removing broken slates. On the right are slate hammers. They have a small striking face and a point on the opposite end for punching nail holes.

of a slate-roof system comes from all its components—the slate itself, the flashings, the fasteners, and the underlayment—acting in concert, and the failure of any component can result in the failure of the whole system and leaks in the living room.

A slate roof is expensive; slate can cost between $250* and $600 per square (100 sq. ft.), and depending on the complexity of the roof, my labor costs can run as high as $450 per square. Given the high cost of the slate, it's foolish to scrimp on the quality of the other components. Using second-rate materials for any part of a slate-roof system is akin to running recapped tires on a Rolls-Royce®.

I use 16-oz. copper for all flashings, copper nails, bituminous membrane at all the eaves, silicone caulk, and 30-lb. felt or roll roofing under the slate. Although some people use standard, galvanized drip edge, a metalsmith makes my drip edge for me out of 16-oz. copper. I have a standard drip-edge profile, which I adjust to fit the house and roof design. It covers at least 1½ in. of the fascia and runs up the roof a minimum of 4 in. I run it along the eaves and up all the rakes. I bend my own step flashings, but there isn't any reason why your tin knocker couldn't make these pieces for you as well.

Order Your Slate Well Ahead of Time

Unless you live near a slate quarry, it's unlikely that you'll be able to go the local roofing supplier and pick up enough slate to do a roof. My regular supplier—I've been using New England Slate for 10 years (1385 U.S. Route 7, Pittsfield, VT 05763; 888-NE-SLATE)—will send me a slate-availability list (along with samples) to match the specification of the job. The list includes an approximate delivery time as well as estimates of delivery costs, via tractor-trailer. My typical slate order takes anywhere from one month to three months to arrive at the job site.

Pipe Staging Holds a Lot of Weight

Because slate is so heavy, loading it on a roof can be a precarious operation. And it seems that no matter how carefully I try to plan things, if I load the roof with more than a couple of days' worth of slate, I spend a lot of time moving the darn stuff around. So I prefer not to stock the whole roof at once. I think it's safer to keep roof surfaces open and free of clutter. And it makes layout easier if the roof isn't full of roofing materials.

It goes without saying that slate is heavy. You need a strong staging on which to stock materials. I like to use steel-pipe scaffolding. It's quick and easy to set up, and although you can rent pipe staging, three months of rental fees about equal the purchase price. If possible, I have a scaffold at all the eaves. I also wear a safety harness when I'm working on a roof (see the sidebar on the facing page).

Thirty-Lb. Felt and Bituminous Membrane Dry-In the Roof

Putting on a slate roof is slow (but satisfying) work. A typical job takes me about three months after the time the slate arrives

Fall Prevention

I think of scaffolding as a way to get materials to the roof, not necessarily as something that keeps me from falling off. To keep from falling, I rely on gear similar to what mountain climbers use (right). I wear a harness that has a sling around each leg and a belt around my waist. A carabiner, which is a big clip kind of like a giant safety pin, attaches to both leg slings and to the belt. I run a series of screw eyes into a rafter every 12 ft. along the roof ridge. A climbing rope is tied to another carabiner and hooked to the screw eye. A smaller rope runs from my harness to the climbing rope. A simple slipknot, the barrel knot, connects the smaller rope to the climbing rope. The barrel knot allows me to move up and down and across the roof while keeping a taut line between me and the ridge.

You'd be surprised how fast you'll get used to wearing the harness and using the ropes. For me, the climbing gear is much more comfortable and infinitely more adjustable than other construction safety belts I have seen. Clerks at a mountaineering store can set you up with everything you need, and they can show you how to use the equipment.

I've been a firm advocate of a rope and climbing harness ever since I was on a job where a man fell 8 ft.—a man who is now a paraplegic.

from the supplier. One huge job I did—170 squares, laid in a graduated, textural pattern—took me a year to complete. Because so much time passes from start to finish and because it's likely that the other trades are finishing off the interior of the house while I'm working, I take special care to dry-in the roof.

Roof dry-in starts with the application of Grace Ice & Water Shield at all eaves (W. R. Grace & Co., www.na.graceconstruction.com). This bituminous membrane is designed to stop water from backing up under the slate with the formation of ice dams, a problem in northern climates for all roofing materials, not just slate. Grace Ice and Water Shield is a self-adhesive flexible membrane, 40mm thick. It's made of polyethylene film and rubberized asphalt. On the back it has a release paper that is peeled off when the membrane is applied to the roof deck.

I lay a 1-ft.-wide strip of membrane on the roof deck starting at the fascia. Over this first piece, I install the copper drip edge. The top 3 in. of the drip edge then are covered by a 3-ft.-wide sheet of membrane. Another sheet goes above this one, lapping the first 3-ft. sheet by 3 in. Two 3-ft. sheets of membrane give me almost 6 ft. of protection from ice damming.

After the second row of bituminous membrane is on the roof, I complete the dry-in with rows of roll roofing or 30-lb. felt. Fifteen-lb. felt is the standard weight for most roofing materials' underlayment. But the additional cost of the heavier materials I use is offset by the fact that they hold up longer without repair.

Roofing nails will hold the paper securely, but I like to use special nails called cap, or button, nails. These nails have a small square of stiff plastic pushed onto the shaft right below the head. The large head of the cap nail holds the larger piece of plastic securely against the felt, effectively resisting the forces of wind and rain.

Two Special Tools Are All You'll Need to Get Started

For around $100, you can get all the special tools needed to put on a slate roof: a slate hammer and a slate cutter. Both are available from New England Slate.

A slate hammer is essential to good slate work. Most slate hammers are lightweight, about 14 oz., and they have a small striking face, usually ¾ in. across. Slate hammers have a long, tapered 6-in. point on the back of the head (see the photo on p. 134). The hammer's point is used for punching nail holes in slate.

Most slate will come from the supplier with a hole punched in each top corner, but there are a lot of times, such as when you cut a large piece of slate into smaller pieces, when you'll have to punch your own holes. It's not hard to do, but like a lot of simple procedures, punching holes takes a little practice.

Because I'm right-handed, I hold the piece of slate face down in my left hand and then give the back of the slate a sharp blow with the point of the slate hammer. The face, or the exposed side of the slate, is the side with the beveled edges. When the slate is cut at the quarry, the shearing action of

A slate hammer pokes a small hole in the back of a slate and makes a beveled hole in the face. A sharp blow with the point of the slate hammer punches a small hole in the back of the slate about 2 in. from the outside edge. The hammer's point pokes a small hole (the size of the hammer point) through the first couple of layers of the slate, and the percussion of the hit blows out the rest of the layers, leaving a conical hole. Ideally, the larger hole on the slate's face makes a countersink for the nail head.

A slate cutter looks like a paper cutter. The author holds the finish side down and uses short, chopping strokes to munch through the piece of slate. The cutter gives a clean cut on the back of the slate and a beveled cut on the face or finish side.

the cutter leaves a beveled edge on one side and a smooth edge on the back. I try to punch the hole about 2 in. from the side edges of the slate and 1 in. more than the exposure line from the bottom. Just the right hit will poke a small hole in the back of the slate (see the top photo on the facing page).

When you turn the piece over, you'll see that the percussion of the blow has blown out a larger hole on the slate's face. The point of the hammer pokes a small hole (the size of the hammer point) through the first couple of layers of the slate (see the bottom left photo on the facing page), and the percussion of the hit blows out the rest of the layers, leaving a conical hole. Ideally, the larger hole on the slate's face makes a countersink for the nail head (see the bottom right photo on the facing page).

For small pieces of slate, say the little pieces of a hip's cap, which a hammer blow might break, I lay the slate face down on a board and punch the hole with the hammer's striking face and a nail set. The less violent blow to the slate sometimes can keep it from breaking.

Several other features also are incorporated into most slate hammers. These features include a nail puller and a slate-cutting edge that is forged into the metal shaft between the handle and the hammerhead.

Some people use the hammer's slate-cutting edge for trimming pieces of slate. I prefer to use a slate cutter, which is similar in design to a paper cutter. The cutting edge is anywhere from 4 in. to 16 in. long.

Slate cutting is a process of shearing, or nibbling, through the slate, rather than a guillotine action. You put the slate in the cutter face-side down. You hold the slate with one hand, and with the other hand you force the cutting edge through the slate in short strokes to nibble away at the stone (see the photo at left). Slate cutting is surprisingly easy, and with practice any intricate shape can be cut.

The slate cutter forces the layers of slate away from the exposed face, and when the stone is turned over, the cut edge is beveled away from the cut. Occasionally, if I am using thick slate or if I am working on a special detail where a square, nonbeveled edge is required, I'll use a wet saw with a diamond blade to make my cuts. For the most part, however, a slate cutter is faster, if not more convenient.

Sheet-Metal Story Poles Speed Layout

Roofing slates are available in different lengths, and the slates' length establishes the weather exposure of the courses on the roof. Longer slates are laid with longer courses. The slates used for the photos in this article are 18 in. long. Other standard lengths are 12 in. and 24 in.

Slate is laid on a roof so that a part of each slates' length is covered by the next two courses above. Because of this overlap, the roof is always three slate layers thick. Slate is laid with a 3-in. head lap. This construction means that the top of each course is covered not only by the next course above but also by the first 3 in. of the second course above. The 3-in. head-lap rule is the basis for

TIP

The best nail for slate work, combining long life and a wide range of sizes, is copper.

Layout tools. The author uses a wax pencil for laying out courses on a sheet-metal story pole. The red wax shows up equally well on slate, sheet metal, and felt paper.

figuring the course height of a certain length of slate. To figure out the course height, take the total length of the slate, subtract 3 in., and divide the difference by two.

For instance, an 18-in. slate minus 3 in. equals 15 in. Dividing that in half leaves a 7½-in. course for an 18-in. slate. Course height can vary by ¾ in. short or ¼ in. long. In order to run an even course up to a ridge or to ensure a course meets, say, the bottom of a skylight, you could run 18-in. slates in courses from 6¾ in. to 7¾ in.

I like to lay out all my slate courses for the whole roof before I nail on the first piece. Establishing my course layout ahead of time prevents confusion when I start laying slate. I lay out my courses on story poles made of 24-ga. galvanized sheet metal that I rip to 2 in. (see the photo above). I pop rivet the strips together in lengths equal to the eaves-to-ridge length of the roof.

To make a matching set of story poles (one for each end of the roof), I nail two strips next to each other at the ridge, letting them hang down the roof to the eaves. With the two strips hanging next to each other, I can mark my course lines on both strips at once. I like to use a mechanical red-wax pencil for marking the story poles (see the photo above). I also use the wax pencil to mark on slate.

Once I have the courses set up on the story poles, I nail one at each edge of the roof and then snap lines, usually six courses at a time. I then roll the story poles to the ridge, where they are out of the way. I first used this system on a complex roof that had seven different slate lengths and different course heights, and since then, I realized that it saves time on even the simplest layout.

Slate Nails Act as Hangers

Slate nails are a critical part of a slate roof's life expectancy. The best nail for slate work, combining long life and a wide range of sizes, is copper. Standard copper slating nails are available in sizes from ½ in. to 3 in. To figure nail length, double the thickness of the slate and add ¾ in. for deck penetration. For this roof I used ⅜-in. slate and 1½ in. nails.

Nailing is the most important skill required for good slate work. A nail must be driven far enough below the surface of the slate (into the countersunk hole made by the hammer) so that the slate above won't rest on the head of the nail and provide a stress-inducing high spot. Conversely, the nail must not pull down so hard on the slate as to break it. The best way to nail a slate is to hold it down snug with one hand and then sneak the nails in just below the surface. Experience will teach you when you've nailed wrong, but only at a price.

Slate nails act more like hangers than fasteners. They are holding the slates on the roof, not holding them to it. If you could pick up a properly nailed slate roof and give it a good shake, the slates would be loose enough to rattle.

The first course is made of three layers of slate. The first and second layers of slate, cut to 4 in. and 10½ in. respectively, are bedded with silicone and nailed even with the drip edge. The third layer is full-length slates. For a hip roof, the full-length slate is notched around the wood strips that provide nailing for the saddle caps.

Triple-Layer First Course Is Bedded in Silicone

In roofing with slate, the first course, starting at the eaves, is three layers thick (see the photo at right). I cut the slate of the first layer to a length of 4 in. I punch holes in these small slates and nail them even with the bottom of the copper drip edge. Before nailing, I bed the slates in two walnut-size blobs of clear silicone caulk. I bed all my small slates—starter courses, ridge caps, small pieces running up a hip—in silicone.

The silicone provides long-term shock protection against things such as someone leaning a ladder against the eaves or walking up a hip. The silicone also provides additional resistance to the unsightly slippage that can sometimes happen with smaller pieces of slate. Few things look worse from the ground than a slate that has slipped below the others in an otherwise straight course on a roof.

The length of the second layer in the first course is determined by adding 3 in. to the full-size course height. If I'm using 18-in. slate, and if the exposure of the courses is 7½ in., then the second layer is 10½ in. long.

The second layer is also bedded in silicone, and it is laid over the 4-in. first layer, even with the drip edge. Next comes my first full-length layer of slate. It covers the first two layers of the starter course and is laid with its bottom edge even with the drip edge. From here, standard coursing, following the lines I've snapped with my story pole, continues to the ridge.

Along with a 3-in. head lap, I lay slate with at least a 3-in. side lap. This layout means that joints between slates in one course should be offset by at least 3 in. from the joints in the course below.

Use Wood Strips for the Ridge and Hip Saddles

I cap all ridges and hips with slate saddles. Each saddle is made of two pieces of 12-in. by 6-in. slate that butt together and extend 6 in. down each side of a ridge or hip. I use a 3-in. head lap for the saddle. Following the same formula used to determine course

A capped hip is six layers thick at the eaves. A three-layer first course is capped by a three-layer saddle cap for a total of six layers of slate.

exposure for the full-length slates gives me a saddle exposure of 4½ in. (12 – 3 = 9; 9/2 = 4½.)

As a base for the saddles, I rip pieces of wood 2 in. wide and as thick as the three overlapping layers of slate that result from the slate course running up the roof. The wood provides a solid base to nail the slate saddle caps to. I nail the pieces of wood along all of the ridges and down all of the hips (see the photo on p. 133), stopping 6 in. from the eave end of the hips and 6 in. from the rakes or gable. If the saddle base pieces went all the way to the eaves, they would be visible from the ground.

Running slate up a hip roof is not difficult, but it does require some careful slate cutting. I make a three-layer first course similar to the straight-eaves first course described earlier. The only real difference is that the pieces have a 45-degree cut along the edge of the hip and a notch cut in the second and third pieces where they go around the wood

Duct tape and string help to align the saddle caps. The pieces of the ridge cap or hip cap (shown here) are bedded in silicone and fastened with two nails each into the wood strips the author attaches to all hips and ridges. To keep the caps straight, he sights along a string that runs from the eaves to the peak. Strips of duct tape hold the two pieces of the caps together while the silicone dries.

saddle base pieces (see the photo on p. 139). After the starter course is run, finishing the hip is a matter of filling in pieces of slate where they meet the solid-wood strips.

At the Eaves, a Hip Is Six Layers Thick

For a hip roof, I start laying the saddle caps at the eaves. Like all first courses, the first course of saddle caps has three layers. The three-layer hip and the three-layer saddle make a total of six layers of slate at the eave end of the hip (see the top photo on the facing page). I punch two nail holes in the upper left-hand or right-hand corners of each saddle piece, depending on which side of the hip they will go on. The holes have to be close together so that the nails will go into the 2-in. wood strip instead of through the slates below (see the bottom photo on the facing page).

As I run the saddle caps up a hip or across a ridge, I bed each piece of cap in silicone. I run a bead of caulk between the two pieces of slate where they meet at the ridge. I stop the bead 4½ in. from the bottom edge so that the caulking won't be visible. Triple coverage of the slates keeps things watertight. Even though every cap is held in place by the two nails in each piece, I use duct tape to hold the saddle slates in place until the silicone dries (see the bottom photo on the facing page). I run a string down the center of a hip or ridge. Aligning the joint where the pieces of each saddle cap meet with the string gives me a straight course.

A Slate Ripper Removes Broken Slates

I always give my customers a limited lifetime guarantee on their slate roofs, but all they have to do to void the guarantee is walk on the roof. Slate is brittle, and it will break easily. If a slate roof is never walked on, the most maintenance you may need to do is replace a couple of slates that inevitably get broken when the roof is installed. When I'm installing a slate roof, I always walk on it gingerly, watching where I step.

Replacing a slate is accomplished using a special tool, a slate ripper. A slate ripper is an 18-in.- to 30-in.-long piece of flat ⅛-in. by 2-in. forged steel attached to a round, offset handle (see the photo on p. 134). A hook is forged on either side of the flat end of the ripper.

To replace a slate, I slide the ripper under the broken slate and use the flat end to locate and then hook over the nails holding the slate in place. Once the nail is hooked, I smack the offset handle of the ripper with a sharp blow from a 2-lb. sledgehammer. A single sharp blow causes less stress to the surrounding slates than repeated light blows.

Once the nails are pulled or cut, the pieces of the broken slate should slide out. A new slate of similar size is bedded in silicone and slid into place. To fasten the new slate, I snip both sides of a nail head to make it T-shaped. I punch a hole in the replaced slate at the upper end of the slot between the slates in the course above. Then I drive the T-headed nail between the slot and through the slate.

Next I fashion a piece of copper to cover the nail in the slot between the slates above. The copper is 3 in. wide and long enough to hook over the top of the replaced slate and cover the T-nail by 3 in. I hem, or bend over, the top ¾ in. of the copper. Then I slip the copper, hemmed-side down, onto the flat end of the ripper and shove the ripper between the replaced slate and the slates above. The hemmed end should snap down past the top of the replaced slate and hook onto the edge. Like all slate-roofing techniques, all this one takes is a little practice.

**Prices noted are from 1995.*

Terry A. Smiley *runs TAS Construction in Woodland Park, Colorado.*

Working Safely on the Roof

■ BY HOWARD STEIN

On his way to a job one morning, a friend of mine stopped to watch a roofing crew climb a single ladder onto a 12-in-12 pitch, two-story roof. There were no scaffolds or roof jacks anywhere in sight. What caught his eye were the two uncovered foam-rubber sofa cushions each roofer carried with him onto the roof. What were the battered cushions for, he wondered? The answer came as he watched each roofer kneel on a cushion and get to work, leapfrogging from cushion to cushion as he shingled upward.

I've worked as a general contractor for nearly 15 years, and I've seen my share of poor safety practices and been guilty of a few myself. Although examples this extreme are rare, there's still plenty of unsafe building to go around. In 1991, for instance, 115,000 construction workers were injured and 158 died as a result of falls.

Although we use standard fall-protection measures such as scaffolds, guardrails on pump jacks and temporary plywood covers over stair and chimney openings in floors, I recently began investing in personal fall-protection equipment for myself and my employees. Before I wrote this article, I experimented with a variety of roof anchors; lifelines, which are safety lines that attach directly to a roof anchor; full-body harnesses; and lanyards, which are short lines that attach harnesses to the anchoring rope. I learned that some types of equipment are easier to use than others and that some are better-made than others. Although the price differences between low-end and better-quality safety equipment can be great, I found that the higher cost usually is worth the investment.

Ignorance of the Law Is No Excuse

Although some contractors may be only vaguely aware of it, the Occupational Safety and Health Administration (OSHA) now monitors residential-construction sites to make sure workers are protected from falls. Changes in OSHA's fall-protection standards consolidate and update existing standards and now also apply specifically to residential construction. Many roofers have complied by purchasing body harnesses, lanyards, rope lines, and anchors, which as a group are called personal fall-arrest systems (see the photo on the facing page). However, general contractors, remodelers, framers,

Free to concentrate. A secure roof anchor, a snug body harness, and a good lifeline allow this roofer to concentrate on the job at hand.

and some of the other trades also are required to comply or face fines up to $70,000, depending on the severity of the violation and the size of the company.

Most of these safety standards aren't new. Many have been in effect for almost 25 years for nonresidential construction. What changed is the recognition that the percentage of homebuilders killed or injured in falls is just as high as that of other construction workers. Those who work in the hazardous construction industry know that falls are the number one cause of serious injury and death. In fact, the revised OSHA rules were generated with the input and support of carpenters' and roofers' unions.

The new fall-protection standards encompass more than just fall-arrest systems. Beyond the scope of this article are other regulations for fall-safety protection, such as the requirement for a written safety plan. I urge you to read the regulations, which are available from OSHA (OSHA Publications, 200 Constitution Ave. NW, Room N3101, Washington, DC, 20210; 202-219-4667). Call your local OSHA office if you need more information.

Three Types of Roof Anchors

With any type of roof anchor, the main point is to secure the worker to the roof. These three types all do the job, but with more or less effort.

Double-strap anchors are versatile and don't require sheathing penetration.

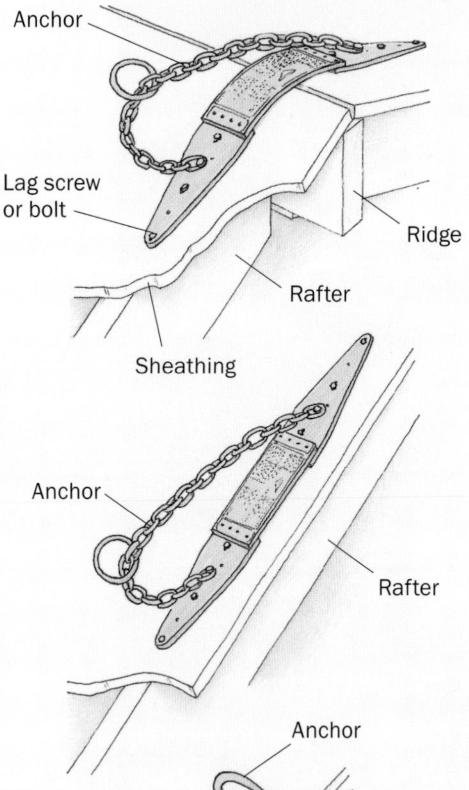

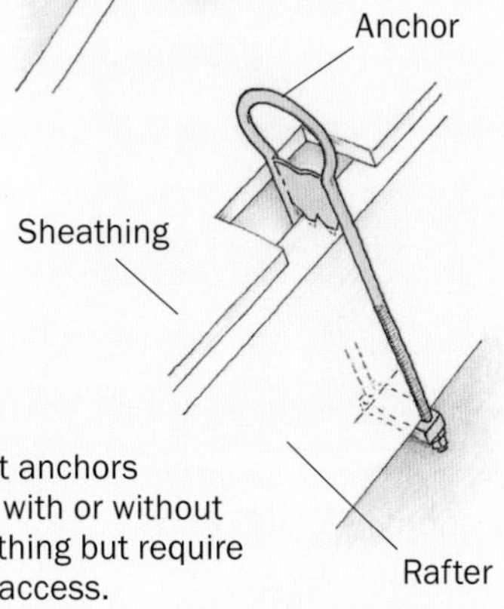

U-bolt anchors work with or without sheathing but require attic access.

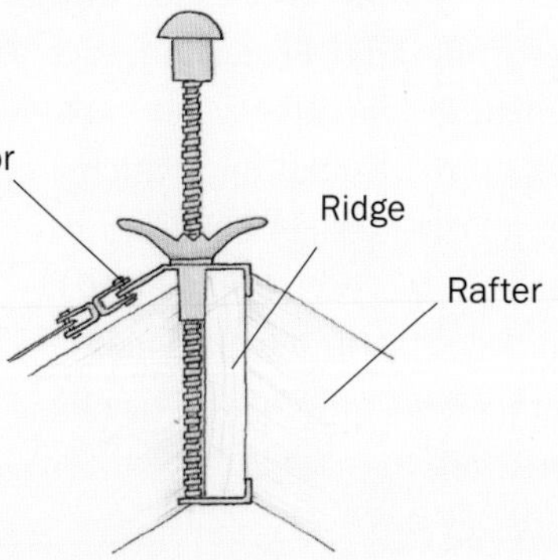

Screw-type anchors fix to ridge or rafter and require a hole in the sheathing.

The Regulations Are for Anybody Working 6 Ft. or Higher in the Air

OSHA's Safety Standards for Fall Protection in the Construction Industry took effect in February 1995 and require some form of fall-protection system at any stage of residential construction where workers are subject to a fall of 6 ft. or more.

Wherever there is an unprotected side or edge at least 6 ft. above a lower level, employees must be protected from a fall by the use of a guardrail, a safety net, or a personal fall-arrest system (PFAS). Work performed from ladders, scaffolds, and pump jacks is subject to its own safety rules. Guardrails and safety-net systems weren't designed for sloped roofs or wood-frame construction.

Homebuilders need to become familiar with the body harnesses, lanyards, rope lines, and anchors that make up personal fall-arrest systems. Manufacturers provide instructions on installation and inspection, cautions on safe use of components and systems, care and maintenance, and application limitations. Instructions are required reading for anyone using this gear.

Safety Starts With a Good Anchor

Whether the following story is true, it illustrates the importance of properly anchoring your lifeline: Crew members working on a roof tie ropes around their waists and heave the lines over the peak of the roof to an apprentice on the ground. The inexperienced lad ties off the ropes to the handiest dead weight around, which happens to be a car bumper. Trusting their anchor point, the crew members begin to work, thinking no more about their security until the car's owner, in a hurry and oblivious to the lifelines, drives away.

A good anchor attaches snugly to the rafter. The author first removed a layer of shingles and drilled a hole in the roof to attach this roof anchor to a rafter. The L-shaped end of the anchor is fed through the hole and secured around the bottom of a rafter. A large wing nut on the threaded shaft is tightened down over a steel plate to fasten the anchor firmly to the roof.

An improvement over car bumpers, roof anchors are attached to the uppermost part of a roof rafter or to the 2x ridge board (see the drawing on the facing page). Manufacturers specify anchor placement about every 8 ft. to 10 ft. and about 6 ft. to 8 ft. from gable ends. These distances are to avoid what they call a swing fall, or "the pendulum effect." The user should be within a 30-degree arc on either side of the anchor point, which on most roofs is within 5 ft. or 6 ft. of the vertical fall line below the roof anchor. Many anchors require that sheathing be in place to help spread the load to other rafters (see the photos above). Double-strap anchors straddle the ridge at any pitch and require filling the field of predrilled heavy-gauge metal straps or steel plates with a specified number of wood screws, nails or lag bolts. These anchors are designed to be attached to rafters through the sheathing, but never into sheathing alone. As a group, these anchors range in price from $6* to $65. The cheaper anchors are disposable or simply bent down and roofed over; the more expensive ones can be used over and over.

Another type of roof anchor can be installed without sheathing in place if 2x stock is nailed above and below the anchor to neighboring rafters. The first 2x is nailed to the top edge of the rafters near the ridge; the second is spiked to the underedges of the rafters below the roof anchor. Ranging in price from $40 to $80, these anchors engage the rafter or ridge by means of a U-bolt or a proprietary design that grabs around the underside of the 2x (see the drawing on the facing page). Each type provides a rugged anchor point on which to connect a lifeline.

You get the kind of rope you pay for. If the only concern is cost, a polypropylene rope (above) will do the job, although it's likely to kink up and cause problems. The braided polyester rope (right) costs a good bit more but will give good service for a lot longer than the cheaper rope.

This kind of anchor can be used at the beginning of roof-sheathing installation. Personally, I wouldn't install them before this stage because until all of the rafters are in place, the roof system isn't stable enough to tie workers off to it and because raftering can be done safely otherwise. Roof anchors are all tested to meet a 5,000-lb. rating and are designed for one person each. The feeling of quality, the ease of use, and the effort to set up and break down the roof anchors varies greatly. After trying out the major types available, I prefer the screw-type Universal Roof anchor from Leading Edge Safety Systems® Inc. (500 Main St., Deep River, CT 06417; 800-241-7330) for temporary use on new work or remodeling and the L3670 stainless-steel, self-flashing anchor from DBI/SALA® (3965 Pepin Ave., Red Wing, MN 55066; 800-328-6146).

Learning the Ropes of Fall-Protection Systems

A PFAS can include two types of rope restraints. Each restrains the worker in case of a fall; both connect directly to the roof anchor. The simpler, less expensive lifeline system starts typically with a 50-ft., ⅝-in.- to ¾-in.-thick synthetic rope (also available in lengths of 30 ft. to 100 ft. or more). The less expensive lines are twisted three-strand rope. The very cheapest of these ropes are polypropylene (see the left photo above) and cost less than $50 for a 60-ft. length. The longest lasting, highest cost lifelines have a braided exterior over a braided core and cost about $135 for the same length. After trying several kinds, I prefer the better braided rope (see the right photo above). A three-strand twisted nylon lifeline (with an eye splice and hook at one end) costs about $75 with a duffel bag for storage. The same in a polyester rope is about $95.

Before I begin a deeper discussion of the various ropes and lines used in fall protection, I need to say something about hardware. A lifeline attaches to the anchor point by means of a self-locking snap hook or carabiner (see the left photo on the facing page). Nonlocking or single-action connector hardware is prohibited after 1997, so most manufacturers only send out hooks and carabiners that require two separate actions to open and that aren't subject to inadvertent rollout of the lifeline, which is when the twisting of the rope from below

It won't open accidentally. This self-locking snap hook, also called a carabiner, is used to attach a lifeline to an anchor and won't open unless two actions are executed.

Rope grabs slide freely but stop sliding under weight. The stainless-steel rope grab connects the lanyard to the lifeline and automatically slides upward when the user ascends. To descend, the worker holds the locking mechanism lever in an upward position so that the cam lock rides freely as he lowers himself. The mechanism grabs the rope if the user slips.

can cause the upper end to roll open the gate on a single-locking safety hook.

A counterweight of virtually any type on the untethered end of the lifeline maintains tension on the rope and allows a worker to slide his tether, or lanyard, up and down the line more easily.

A piece of stainless-steel hardware called a rope grab connects the lanyard to the lifeline (see the right photo above). On the other end, the lanyard attaches to a D-ring on the body harness with a double-locking hook or carabiner. All rope grabs automatically slide upward if the user ascends; some brands move more freely than others. My favorite rope grab is made by SURETY Manufacturing & Testing Ltd. (2115 91st Ave., Edmonton, Alberta, Canada T6P 1L1; 800-661-3013). To descend, the worker holds the locking mechanism lever in an upward position so that the cam lock rides freely as he lowers himself by holding the lifeline with his other hand. In case of a slip, the mechanism grabs the rope. The mechanism allows, on demand, a free-sliding up-and-down action for hands-free use. It costs about twice the average grab.

Rope grabs are tested for use with a particular lifeline and lanyard and shouldn't be mixed with components from other suppliers without first checking for compatibility.

Here's a better way to hook lanyard to harness. It's always awkward and often difficult to reach behind the back and hook a lanyard to the D-ring that's attached to the body harness. This way works better. A 1-ft. lanyard that contains a shock absorber is factory-attached to the back of the harness, which can be swung around for attachment to a lanyard.

Suiting Up With Full-Body Harnesses

At the opposite end of the lifeline from the anchor point is the full-body harness, worn by each worker as part of the PFAS. Harnesses are available in all sizes and many styles, and they adjust by means of belt and buckle, parachute-type friction buckle or pass-through bar buckle. Some manufacturers offer only a universal size.

The D-ring on a fall-arrest harness is on the back, where the rear webbing crosses between the shoulder blades. Like all other components of the system, the D-ring is rated at 5,000 lb. All have straps that adjust around the thighs just below the buttocks to distribute the impact of a fall.

The harness should fit well and be adjusted properly for comfort and safety. Harness prices range from about $60 to $130. There are many good-quality harnesses in the $65 to $80 range.

The Proper Lanyard Is No Longer than 3 Ft.

The lanyard that connects the rope-grab mechanism to the lifeline is available in lengths from 12 in. to 36 in., but in no case should it be longer than that. Lanyards are made of synthetic webbing or rope and come with or without a built-in shock, or energy, absorber. The shock absorber is usually a sealed, rip-stitch section of folded webbing that gives up its slack when subjected to the stress of a fall.

On every piece of equipment I've tested, the lanyard is factory-attached to the rope grab, which means users have to reach over their backs to connect the lanyard to the D-ring on the back of the harness. This construction can be frustrating, especially if you're wearing gloves, because the double-locking snap hook or carabiner requires the two separate actions to engage or disengage.

Both Prevent Falls

The self-retracting lifeline, left, works like a seat belt and allows hands-free movement but costs several times more than the rope-grab system.

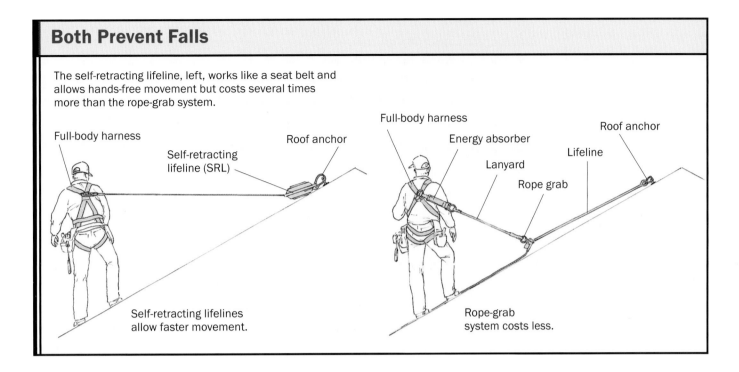

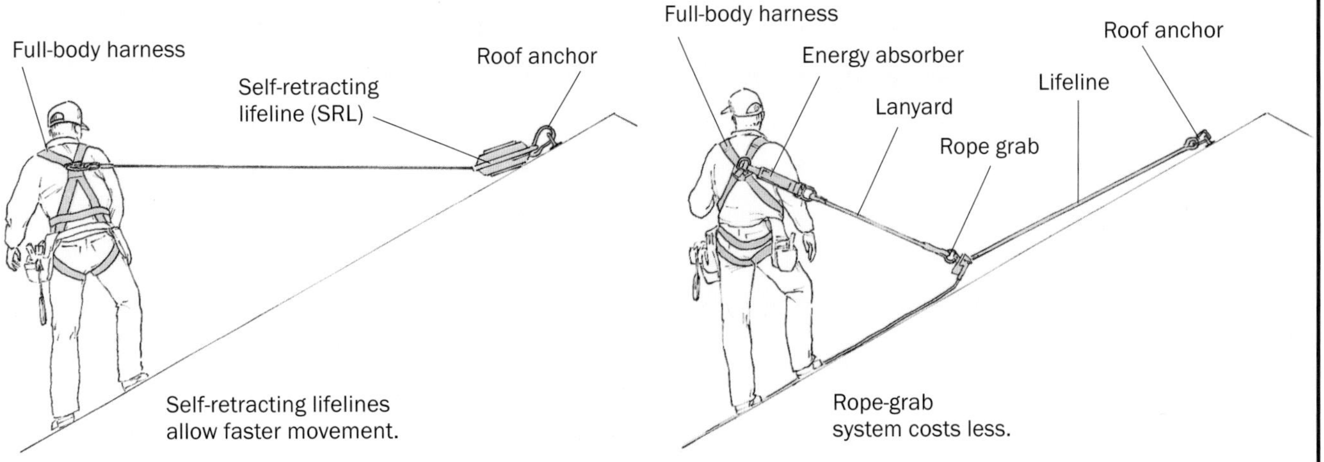

I found the best combination to be a 1-ft.-long shock absorber that's factory-connected to the harness, and a separate 1-ft. lanyard attached to the rope grab. The two hook together easily without requiring you to reach behind your back (see the right photo on p. 147 and the photo on the facing page). For me, a lanyard of 18 in. to 24 in. is the most comfortable; a 36-in. lanyard requires a step up the slope to reach and maneuver the grab.

As a caution, there are systems on the market that come with a 6-ft. lanyard that meets standards for strength but that allows up to a 12-ft. fall if the user gets the full length of the lanyard above the rope grab and forgets to take up the slack. However, a 6-ft. lanyard with hook at each end is useful for working a single area near a roof anchor for tasks such as repointing a chimney. If this location is near an edge—less than 6 ft. from a gable end, for example—it should include an integral shock absorber.

Dennis Garafolo of Sinco Inc. (701 Middle St., Middletown, CT 06457; 800-243-6753) warns that although his company sells many of the less expensive lifeline packages to contractors who want to get into quick compliance with the OSHA rules, most workers use rope-grab systems incorrectly. "A worker will snap onto an anchor at the ridge, walk down to the eave, lock on the grab, and then move up and down the roof without adjusting the rope grab to take up the slack as he moves around the roof. This can develop a 20-ft. lanyard, which is not user-friendly." In other words, it's dangerous.

Self-Retracting Lifelines Are More Expensive but Easier to Use

A self-retracting lifeline (SRL) system automatically retracts or pays out cable from a metal or high-impact plastic housing as a worker moves toward or away from the unit (see the left drawing above). Inside is a speed-sensing brake, much as in a seat belt, that activates in a free-fall and cuts the impact force on the user.

SRLs are heavy, so they mount directly to the anchor point. A carabiner or hook at the end of the cable attaches to the harness. It's important to specify an SRL that's made to operate on a slope because some are designed for vertical use only and can bind when used on a roof pitch.

An SRL with 50 ft. of cable and a large carabiner costs about $850. A cheaper SRL option has a thermoplastic housing and 20-ft., 1-in.-wide synthetic webbing instead of cable. It costs about $570. For comparison, a 50-ft. lifeline and rope grab with attached 18-in. or 24-in. lanyard cost from $160 to $300, depending on their features and quality.

The Safety Package You Use May Depend on Your Trade

Just as a roofer uses a different type of hammer than a house framer or a finish carpenter, components of personal fall-arrest systems can vary from trade to trade. What works best for one trade might not be the best system for another.

A general contractor could supply roof anchors on a job, carrying the cost just as he does with a portable toilet or a job-site fire extinguisher. That way, all of the other trades could tie off their equipment to his anchors. (As a general contractor, I feel comfortable sharing roof anchors, but not lifelines and harnesses.)

- Roofers move around faster than carpenters do. To maintain production, many roofing contractors equip themselves with SRLs. Although more expensive, SRLs give quicker payback because they increase efficiency every day.

  Some roofers' packages include a waist belt with a rope-grab lifeline system. Belts aren't acceptable after 1997, except on low-slope roofs (such as 4-in-12 pitch or less), because a free-fall in a belt can cause serious back injury. A majority of roofers have switched to full-body harnesses for that reason.
- Carpenters spend less time on the roof as a percentage of total work and, when sheathing a roof, don't need to move horizontally or vertically as fast or as often as roofers. For a two-person roof-sheathing crew aided by a cut person below, a combination of systems might be a good investment. The more-experienced carpenter would use an SRL, which leaves no rope at his feet to trip on as he focuses on nailing off the ply. The other person is tied off on a less-expensive lifeline with rope grab. This system remains in place for trimming the roof eaves with fascia and soffit boards; the crew is tied off while working from planks on sidewall brackets at the second- and third-story level. If the roofing crew shares the roof anchors, only one setup is needed.

There's a Bottom Line to Safe Building

The time it takes to set up, use, and break down PFASs could make a contractor slightly less competitive against crews that don't comply with OSHA's regulations—unless the noncomplying crews get caught by OSHA. But when all contractors begin to comply with the rules, those builders who started early will already have a competitive edge.

For one thing, a good safety record keeps worker's compensation premiums lower. The mandatory accident report to the insurance company could trigger an investigation about whether fall protection was in use. Besides the obvious emotional issues involved, a serious injury or death could financially devastate a contractor. Accidents also disrupt work, and they can involve the costs of both time and money to train replacement workers.

There are other benefits to using a PFAS. A one-person crew alone on a site is much safer using one of these systems. And some tasks are performed faster because the worker's efforts aren't focused on avoiding a fall. Also, on steeper roofs, lifelines are helpful in climbing. Finally, PFASs might restrict movement but can also be used for temporary positioning to avoid a lengthy setup for a short task.

**Prices noted are from 1996.*

***Howard Stein** is a contractor in West Townsend, Massachusetts.*

CREDITS

The articles in this book appeared in the following issues of *Fine Homebuilding:*

p. iii: Photo by Roe A. Osborne, courtesy *Fine Homebuilding*, © The Taunton Press, Inc.; p. iv: Photo by Tom O'Brien, courtesy *Fine Homebuilding*, © The Taunton Press, Inc.; p. v: Photo by Roe A. Osborne, courtesy *Fine Homebuilding*, © The Taunton Press, Inc.

p. 4: Flashing Walls by Scott McBride, issue 100. Photos by Scott McBride, except for p. 4 by Andrew Wormer, courtesy *Fine Homebuilding*, © The Taunton Press, Inc; Illustrations by Dan Thornton, courtesy *Fine Homebuilding*, © The Taunton Press, Inc.

p. 14: How to Avoid Common Flashing Errors by James R. Larson, issue 115. Photo by Scott Gibson, courtesy *Fine Homebuilding*, © The Taunton Press, Inc.; Illustrations by Christopher Clapp, courtesy *Fine Homebuilding*, © The Taunton Press, Inc.

p. 22: All About Rain Gutters by Andy Engel, issue 125. Photos by Andy Engel, courtesy of *Fine Homebuilding*, © The Taunton Press, Inc.; Illustrations by Dan Thornton, courtesy *Fine Homebuilding*, © The Taunton Press, Inc.

p. 31: Draining Gutter Runoff by Byron Papa, issue 107. Photos by Byron Papa; Illustrations by Chuck Lockhart, courtesy *Fine Homebuilding*, © The Taunton Press, Inc.

p. 34: Roof Flashing by Scott McBride, issue 118. Photos by Scott McBride, courtesy *Fine Homebuilding*, © The Taunton Press, Inc.; Illustrations by Chuck Lockhart, courtesy *Fine Homebuilding*, © The Taunton Press, Inc.

p. 43: Flashing a Chimney by John Carroll, issue 136. Photos by Tom O'Brien, courtesy *Fine Homebuilding*, © The Taunton Press, Inc.; Illustrations by Chuck Lockhart, courtesy *Fine Homebuilding*, © The Taunton Press, Inc.

p. 51: Preventing Ice Dams by Paul Fisette, issue 98. Photos by Paul Fisette; Illustrations by Dan Thornton, courtesy *Fine Homebuilding*, © The Taunton Press, Inc.

p. 58: Four Ways to Shingle a Valley by Mike Guertin, issue 152. Photos by Roe A. Osborne, courtesy *Fine Homebuilding*, © The Taunton Press, Inc.

p. 69: Laying Out Three-Tab Shingles by John Carroll, issue 90. Photo by Jefferson Kolle, courtesy *Fine Homebuilding*, © The Taunton Press, Inc.; Illustrations by Christopher Clapp, courtesy *Fine Homebuilding*, © The Taunton Press, Inc.

p. 77: Tearing Off Old Roofing by Jack LeVert, issue 86. Photos by Rich Ziegner, courtesy *Fine Homebuilding*, © The Taunton Press, Inc.

p. 87: Reroofing With Asphalt Shingles by Stephen Hazlett, issue 138. Photos by David Ericson; Illustrations by Christopher Clapp, courtesy *Fine Homebuilding*, © The Taunton Press, Inc.

p. 95: Aligning Eaves on Irregular Pitched Roofs by Scott McBride, issue 91. Photos by Scott McBride; Illustrations by Bob Goodfellow,© The Taunton Press, Inc.

p. 102: Installing a Rubber Roof by Rick Arnold and Mike Guertin, issue 113. Photos by Photos by Roe A. Osborne, courtesy *Fine Homebuilding*, © The Taunton Press, Inc. except for photo on p. 111 by Rick Arnold and Mike Guertin; Illustrations by Dan Thornton, courtesy *Fine Homebuilding*, © The Taunton Press, Inc.

p. 112: Installing Steel Roofing by John La Torre Jr., issue 134. Photos by Photos by Roe A. Osborne, courtesy *Fine Homebuilding*, © The Taunton Press, Inc.; Illustrations by Chuck Lockhart, courtesy *Fine Homebuilding*, © The Taunton Press, Inc.

p. 122: Choosing Roofing by Jefferson Kolle, issue 92. Photos: p. 123 and p. 125 (top) courtesy Certainteed Corporation; p. 125 (bottom) and 126 courtesy Cedar Shake & Shingle Bureau; p. 127 courtesy Met-Tile Inc.; p. 128 courtesy Alcoa Building Products; p. 129 courtesy Bethlehem Steel; p. 129 courtesy M.C.A. Inc.; p. 131 (top) by Terry Smiley (bottom) courtesy Certainteed Corporation; p. 132 by Terry Smiley.

p. 133: Roofing With Slate by Terry A. Smiley, issue 96. Photos by Jefferson Kolle, courtesy *Fine Homebuilding*, © The Taunton Press, Inc.

p. 142: Working Safely on the Roof by Howard Stein, issue 99. Photos by Steve Culpepper, courtesy *Fine Homebuilding*, © The Taunton Press, Inc.; Illustrations by Vince Babak, courtesy *Fine Homebuilding*, © The Taunton Press, Inc.

INDEX

T

U

V

W

Z

Taunton's FOR PROS BY PROS Series
A collection of the best articles from *Fine Homebuilding* magazine.

Other Books in the Series:

Taunton's For Pros By Pros:
RENOVATING A BATHROOM
ISBN 1-56158-584-X
Product #070702
$17.95 U.S.
$25.95 Canada

Taunton's For Pros By Pros:
BUILDING ADDITIONS
ISBN 1-56158-699-4
Product #070779
$17.95 U.S.
$25.95 Canada

Taunton's For Pros By Pros:
BUILDING STAIRS
ISBN 1-56158-653-6
Product #070742
$17.95 U.S.
$25.95 Canada

Taunton's For Pros By Pros:
BUILT-INS AND STORAGE
ISBN 1-56158-700-1
Product #070780
$17.95 U.S.
$25.95 Canada

Taunton's For Pros By Pros:
EXTERIOR SIDING, TRIM & FINISHES
ISBN 1-56158-652-8
Product #070741
$17.95 U.S.
$25.95 Canada

Taunton's For Pros By Pros:
FINISH CARPENTRY
ISBN 1-56158-536-X
Product #070633
$17.95 U.S.
$25.95 Canada

Taunton's For Pros By Pros:
FOUNDATIONS AND CONCRETE WORK
ISBN 1-56158-537-8
Product #070635
$17.95 U.S.
$25.95 Canada

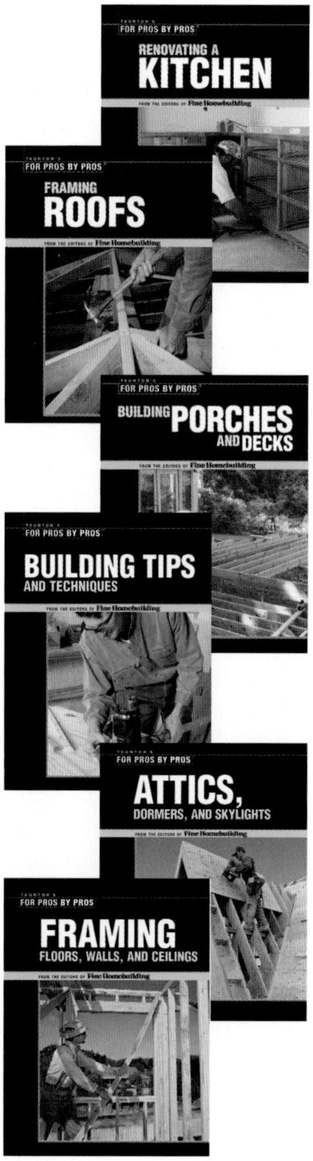

Taunton's For Pros By Pr
RENOVATING A KITCH
ISBN 1-56158-540-8
Product #070637
$17.95 U.S.
$25.95 Canada

Taunton's For Pros By Pr
FRAMING ROOFS
ISBN 1-56158-538-6
Product #070634
$17.95 U.S.
$25.95 Canada

Taunton's For Pros By Pr
BUILDING PORCHES A DECKS
ISBN 1-56158-539-4
Product #070636
$17.95 U.S.
$25.95 Canada

Taunton's For Pros By Pr
BUILDING TIPS AND TECHNIQUES
ISBN 1-56158-687-0
Product #070766
$17.95 U.S.
$25.95 Canada

Taunton's For Pros By Pros
ATTICS, DORMERS, AN SKYLIGHTS
ISBN 1-56158-779-6
Product #070834
$17.95 U.S.
$25.95 Canada

Taunton's For Pros By Pros
FRAMING FLOORS, WALLS, AND CEILINGS
ISBN 1-56158-758-3
Product #070821
$17.95 U.S.
$25.95 Canada

For more information visit our website at www.taunton.com.